YOUR GUIDE
TO LEAN DESIGN & CONSTRUCTION

WITH GLOSSARY COURTESY OF
THE LEAN CONSTRUCTION INSTITUTE

JULIE GLASSMEYER
1st Edition

Copyright © 2022 by Julie Glassmeyer

All rights reserved. No part of this book may be reproduced or used in any manner without written permission of the copyright owner except for the use of quotations in a book review. For more information, address: YourGuide@glassmeyerconsulting.com.

First paperback edition August 2022

Book design by Michele Messing and Charlie McGroarty
Illustrations by Charlie McGroarty

ISBN 979-8-9862413-0-2 (paperback)
ISBN 979-8-9862413-1-9 (e-book)

Published by IngramSpark
www.glassmeyerconsulting.com

"

OPERATIONS KEEPS THE LIGHTS ON, STRATEGY PROVIDES A LIGHT AT THE END OF THE TUNNEL, BUT PROJECT MANAGEMENT IS THE TRAIN ENGINE THAT MOVES THE ORGANIZATION FORWARD.

-JOY GUMZ

"

CONTENT

ALLOW ME TO MENTION SOME AMAZING PEOPLE

I have been extremely fortunate to learn and work with numerous smart, innovative, and helpful individuals and organizations, throughout my career and along my Lean Journey, without whom this book would never have come to be. I would be remiss if I didn't personally thank each and every one of them... Yeah, I don't have enough pages, so let me at least try to hit some of the biggies.

First, and foremost, I need to thank the man that lured me into the wild world of Lean coaching, Dave MacNeel (On Point Lean). I count Dave as a friend, colleague, and mentor – since years before I became a coach – not to mention the best consultant a sub-consultant could ever ask for.

Thank you to Messer Construction Company, and specifically Steve Keckeis, Mark Luegering, and Andy Burg, for introducing me to Lean Construction so many years ago, giving me the opportunity to implement and learn as I went along as a Project Manager and Senior Project Manager, and coaching me through every early struggle.

My former work family at Grote Enterprises gave me the opportunity to spread my Lean wings and allowed me to really grow my training, coaching, and facilitating skills. That was also the place I honed my ability to effectively articulate the principles of Lean Design and Construction in writing.

Now, let's talk about actually getting this book written, designed, and published. I can never express how much I appreciate the assistance Michele Messing and Charlie McGroarty provided for the cover design and interior design (Actually, they didn't assist. They did it all!). Charlie also brought my imaginary project team to life with his illustrations. Kristin Hill (Lean Construction Institute), Joe Newton (Turner Construction Company), and Alla Reese (Key Performance Skills) all volunteered their personal time to read multiple drafts and provide excellent feedback and constructive criticism. I can never forget David Crawford (InView Group) for suggesting I write a book in the first place. (Sorry it took so long, David!) To my friends and family that listened to me talk about this book for about a year (maybe more), I thank you for your love, patience, and encouragement.

To round out this seemingly, yet not even close to, exhaustive list of people to whom I am ever so grateful, big thanks go out to the Lean Construction Institute – national, the Ohio Valley Community of Practice, and the Lean Coaches Community of Practice - as well as every team I've ever coached. I have learned so much from each and every one of you!

SOME THOUGHTS FROM THE AUTHOR

Thank you for taking the time to read this guide to Lean project implementation. I have really enjoyed developing this book, and I hope it provides you with the tools you'll need to be a successful participant on a Lean project.

Before we get started, I want to give you some thoughts to keep in mind while reading the book and all along your Lean Journey.

1. **The instructions in this guide are a compilation of best practices that I've collected over my two decades of learning, implementing, and coaching Lean Design & Construction principles.** They have evolved, and, I imagine, will continue to evolve. I have been privileged to learn from and work with some very smart and experienced Lean practitioners over the years. Take a look at the Acknowledgements for a (by no means all-inclusive) list of those people.

2. **That being said, you should go into this journey with the understanding that everyone does Lean a little differently.** You will experience these differences from person-to-person, company-to-company, and project-to-project, and that's OK. As long as the tools and their purpose remain the same, some deviations in implementation are to be expected. This guide describes a simple plan for implementation, along with explanations of the purpose and outcome for each step along the way, so you and your team can stay focused on the core principles and achieve the benefits of Lean Design & Construction.

3. **Enjoy the journey!** Beginning a Lean initiative, whether it be a company, project, or individual, isn't easy. It takes planning, discipline, and a willingness to break out of your traditional mindset, but it can also be fun. The whole point is to break down barriers, improve processes and outcomes, and

share responsibility with an aligned and collaborative team. Be optimistic. Be bold. Trust each other. And please, please, please celebrate your successes, both big and small!

4. **Be kind to yourself and your team.** As familiar as some of the concepts and tools may feel, Lean is different from the way most of us were brought up to view and manage design and construction. It takes dedication and effort to remain consistent and true to the principles described in this book. You and your team will probably backslide, deviate, and revert to more traditional methods on occasion. If you remain vigilant and committed, you can easily recognize a divergence from the plan and get back on the Lean path.

> **[A LEAN INITIATIVE] TAKES PLANNING, DISCIPLINE, AND A WILLINGNESS TO BREAK OUT OF YOUR TRADITIONAL MINDSET, BUT IT CAN ALSO BE FUN.**
>
> —————— Julie Glassmeyer

1

"ART IS THE ELIMINATION OF THE UNNECESSARY."

-Pablo Picasso

CHAPTER ONE

WELCOME

WELCOME TO THE WORLD OF LEAN DESIGN & CONSTRUCTION!

Whether you're a professional about to embark on your Lean Journey, a student getting your first taste of Lean, or an experienced Lean practitioner seeking additional guidance for yourself and your team, congratulations on taking this step to further our industry's ability to boost project teamwork and collaboration, grow a culture of trust and respect, and optimize project outcomes for everyone involved.

You may be asking yourself, "Why Lean? Who does it benefit?" The answer is: Every participant on the project! According to two studies performed in 2016 – "Why Do Projects Excel? The Business Case for Lean Project Delivery," performed by Dodge Data Analytics, and "Motivation and Means: How and Why IPD and Lean Lead to Success," performed by the University of Minnesota – projects that participate in high intensity Lean are three times more likely to complete ahead of schedule and almost twice as likely to finish below budget, when compared to low intensity Lean projects.

This guide has been designed to describe a general process/toolkit/ delivery system/etc. that can be implemented during design and construction. Along with training, coaching, and sustained effort by all team members, the information enclosed will support efforts to quickly create and maintain a high performing team focused on innovation, continuous improvement, eliminating waste, and optimizing value.

Please take the time to read this book thoroughly, highlighting the topics that are most important to your specific role, and keep it handy for reference.

LEARNING OBJECTIVES:

- Have a foundational understanding of Lean Design & Construction concepts and goals;

- Understand, and be able to describe, the purpose and benefits of employing Last Planner System® tools and processes;

- Be prepared to effectively participate in a Last Planner System implementation;

- Understand, and be able to describe, the purpose and benefits of utilizing Target Value Delivery;

- Be prepared to effectively participate on a Target Value Delivery project team; and

- Have a resource when questions arise.

2

> "IF WE CAN FALL IN LOVE WITH SERVING PEOPLE, CREATING VALUE, SOLVING PROBLEMS, BUILDING VALUABLE CONNECTIONS AND DOING WORK THAT MATTERS, IT MAKES IT FAR MORE LIKELY WE'RE GOING TO DO IMPORTANT WORK."

-Seth Godin

CHAPTER TWO

LEAN OVERVIEW

Lean Design & Construction is a project management philosophy derived from the Toyota Production System. A project team committed to the tenets of Lean focus intently on the identification and elimination of waste in the design and construction process, delivery of value as defined by the customer, respect for people, continuous improvement, planning for flow, deliberate collaboration, and problem solving by those performing the work.

WASTE IDENTIFICATION & ELIMINATION

It is widely accepted that over 50% of time spent in design and construction can be classified as "waste." While that may be a daunting percentage, that number offers great opportunity for any project team committed to improvement. If there is that much waste in our processes, it would only take small improvements to begin moving the needle in a better direction. The first step toward removing waste is identifying it. To that end, Toyota defined "7 Deadly Wastes."

1. *TRANSPORTATION – Excess travel distance. This can refer to travel to or around the worksite.*

2. *INVENTORY – Excess materials purchased and stored.*

3. *MOTION – Excess movement of the material or operator.*

4. *WAIT – Material or operator in idle time.*

5. *OVER PRODUCTION – Made too much or performed the work too early.*

6. *OVER PROCESSING – Unnecessary or repeated steps.*

7. *DEFECTS – Doesn't meet requirements or requires rework.*

Further, upon adaptation to the design and construction industry, we have added two additional wastes that damage our ability to provide value to our customers.

8. *UNDERUTILIZED CREATIVITY OF TEAM MEMBERS*

9. *WORK-AROUNDS – FIREFIGHTING – MAKING DO*

DELIVERY OF VALUE AS DEFINED BY THE CUSTOMER

In the design and construction industry, we have a habit of assuming we know the meaning of a "successful project." We believe that the four pillars of success are QUALITY, SAFETY, COST, and SCHEDULE.

Imagine if you were building a casino, though. Specifically, consider the Encore Boston Harbor Casino, which brought in $16.8 million in their first week of business (*Boston Globe* 7/15/19, Jon Chesto). If you were the owner of that casino, would cost be your biggest driver or would schedule? Let's be honest. They would have happily paid an extra $500,000 to get that casino opened a week early.

Now, imagine building a community center in urban Detroit. We can assume that cost is going to be a driving force for this customer. However, they may also be committed to creating a safe space for the neighborhood youth, engaging the surrounding area during construction, and inspiring outside private investment.

How can we, as designers and constructors, possibly know what constitutes a successful project for our customers if we don't ask the question? How can we deliver a successful project if we don't know what that means? We need to ask the question.

RESPECT FOR PEOPLE

All the planning in the world is useless without the human capital to make it a reality. With this in mind, it is imperative that we respect all people that are a part of making our projects come to life – owner, designers, contractors, end users, and labor force. We must respect their contributions, suggestions, opinions, and time.

When organizing a Lean project, the leadership team must consider the best methods for soliciting and utilizing the input of every member of the team. Build in opportunities for retrospective analysis at the end of every meeting and planning session throughout the project to determine your progressive success at meeting project goals, and upon project completion to capture and share lessons learned.

Challenge yourselves to open the planning process to Last Planners (foremen, superintendents, and design leads) that understand the design and construction process for their individual trades and disciplines better than anyone else could. By implementing Pull Plans, Daily Huddles, and Weekly Work Plans (see Chapter 3 for the Last Planner System® Tools), we engage these experts to develop and carry out detailed project plans with better results than we've ever achieved through traditional project management techniques.

Finally, by taking the time to understand the Conditions of Satisfaction for all team members, we allow for a reliable flow of work through the design and construction process and into building operation. "Conditions of Satisfaction" refers to the specific quality/ characteristics of work required, upon handoff, for the next worker/ user in the supply chain to do their work successfully. This may mean that the drywall needs to be touched up and the area swept for the painter to do their work, or it could mean that the building is fully operational and punched-out for the veterinary clinic to begin their day-to-day operations. Either way, we need to provide a forum for Conditions of Satisfaction to be communicated, in order to keep the project flowing toward completion.

CONTINUOUS IMPROVEMENT

Always pursue perfection. We are human, and we will never be perfect. The lesson we take from that statement is that there is ALWAYS room for improvement. If we take a process that used to take 2 hours and make it take 1.5 hours, there is still opportunity to make it take 1 hour. If a typical school project costs $30 million, we can make the next one cost $25 million. As long as we recognize the infinite possibilities for improvement, we can always strive to achieve it.

If we, as a team, challenge ourselves to do better on every project, we can do anything. This commitment to continuous improvement will lead to creative tension, requiring collaboration and innovation at every turn. Begin every project with a planning session with all design and/or trade partners to flush out concerns and big ideas.. Identify the ideas with the highest likelihood of delivering value to the project, and pursue those ideas with passion. Where will your team go – prefab, modular, kitting, big room, Integrated Project Delivery, logistics planning, just-in-time deliveries, shared services, prototyping, virtual reality? The possibilities are endless!

Outside of the big improvements that we can achieve for our overall project outcomes are the sometimes small improvements we can make in the experience for our team members. How can we make coming to work in the morning a less daunting endeavor for the individuals and crews that help us reach our goals? To do this, we need to be open to suggestions and complaints, and take action to improve the quality of life for everyone. This means asking for critical feedback and suggestions for improvement... and then doing something about it. Nothing leads to better productivity improvement like caring about the needs of the people doing the work.

PLANNING FOR FLOW

Flow is everything. In every aspect of our lives, our experiences, productivity, and attitudes are improved when we find ways to increase the level of flow in the things we do – driving to work, moving from terminal to terminal at the airport, progressing through the activities on a jobsite. If that's true, why do we do so many things that impede flow – tailgating on a busy highway, blocking others from walking up and down the escalator, allowing an aggressive contractor to "push" the others? The ugly fact is that, when we are behaving in an aggressive or urgent manner, we feel like we're getting more done.

The truth is just the opposite. On our design and construction projects, we should be focusing on the entire value stream of the product we are creating and working as a team to ensure reliable handoffs of work from one trade/discipline to the next. By working with our discipline leads / trade partners to right-size the work, build in achievable durations, and enable commitments between the Last Planners, we can develop a project workflow that results in a shorter overall duration.

Now, here is the danger of working on a project that has achieved a reliable workflow. It feels slow. We are used to the chaos of a traditional construction project. We expect conflict between our Last Planners, work to be done out of sequence, damage to materials and finished work, and a mad dash to the finish line. But, if you collaborate as a team, plan ahead, identify and eliminate constraints well in advance, and communicate constantly, these hallmarks of a typical project are drastically reduced. Hence the feeling of slowness.

What you're actually feeling is a calm and productive working environment. When your project is progressing calmly and reliably, the leadership team and Last Planners can put their efforts towards planning upcoming work and looking for ways to innovate and further improve the process.

Flow is everything.

DELIBERATE COLLABORATION

Collaboration is the key to achieving the goals of Lean design and construction. We must communicate and work together to identify and eliminate waste, improve our processes, and provide the greatest value! In every industry, you will meet truly collaborative people. They seek out input from others, facilitate open discussions, and look for consensus on the best way to tackle challenges. Design and construction are no different, and you have probably been fortunate enough to work with some of these people. However, our projects are populated by many very talented individuals with varying personalities and communication styles. For that reason, it is imperative that we make a deliberate effort to create an environment that both encourages and enables collaboration between our team members. This means planning. What regular big room / cluster group / planning / huddle meetings are required to ensure that we can reliably progress toward completion / identify obstacles for resolution / innovate to improve on our processes? We cannot assume that collaboration will happen organically. We must create a plan, stick to the plan, and make adjustments as necessary.

PROBLEM SOLVING BY THOSE PERFORMING THE WORK

We will delve further into this concept in the Last Planner System chapters, but it's important to identify this as one of the guiding concepts of our Lean initiative. The "Last Planners" are the individuals with final control over the work that is actually being performed. In design this would be the discipline leads, and in construction it is the foremen and site superintendents. Traditionally, work assignments have been pushed down to these Last Planners from managers more removed from the actual work. This often leads to work being performed in a less than ideal sequence, rework, and jumping around. Last Planners have very little control over their ability to actually perform the planned work, and therefore, justifiably step further away from the planning process. Commitments are unreliable, conflict is common, and productivity is low.

By shifting responsibility for planning to the Last Planners and empowering them to identify obstacles to their progress for removal by project leadership, the Last Planners can focus on providing value every day. Imagine a project where all of the Last Planners understand the overall plan and their part in it, because they devised it. Schedule reliability and flow increase. Communication between Last Planners is eased, because they are collaborating regularly. Safety and quality improve, and waste is reduced, because work is being planned ahead of time. In general, the atmosphere of a Lean project is much different than a traditional project.

3

"ONE OF THE MOST SINCERE FORMS OF RESPECT IS ACTUALLY LISTENING TO WHAT ANOTHER HAS TO SAY."

-Bryant H. McGill

CHAPTER THREE

LAST PLANNER SYSTEM® OVERVIEW - TOOLS

So, now you've been introduced to Lean Design & Construction. Lean is the overarching philosophy that describes the behaviors that we are striving to achieve. The Last Planner System® (LPS) is a kit of tools that will enable the entire team to behave in a Lean manner. In this chapter, you will find an overview of LPS tools, and how teams use them to deliver a Lean project. In the next chapter, we'll talk about the meetings that support a Lean implementation.

As we discussed in the previous chapter, the "Last Planners" are the individuals that have final control over the work that is being put in place (Design – discipline leads; Construction – foremen & site superintendents). The goal of LPS is to empower the Last Planners to plan their work together, leading to more reliable and flexible plans and workflow, faster problem solving, better communication, and enhanced value delivery.

The tools of the Last Planner System include:

MILESTONE PLANNING/MASTER PLANNING
PHASE PULL PLANNING
LOOK-AHEAD PLANNING
WEEKLY WORK PLANNING
PERCENT PLAN COMPLETE/VARIANCE TRACKING

The system is supported by two regular meetings (+ one prep meeting that improves the efficiency of the Weekly Work Plan Meeting):

WEEKLY WORK PLAN MEETING
DAILY HUDDLE/STAND-UP
LEADERSHIP TEAM WWP MEETING PREP

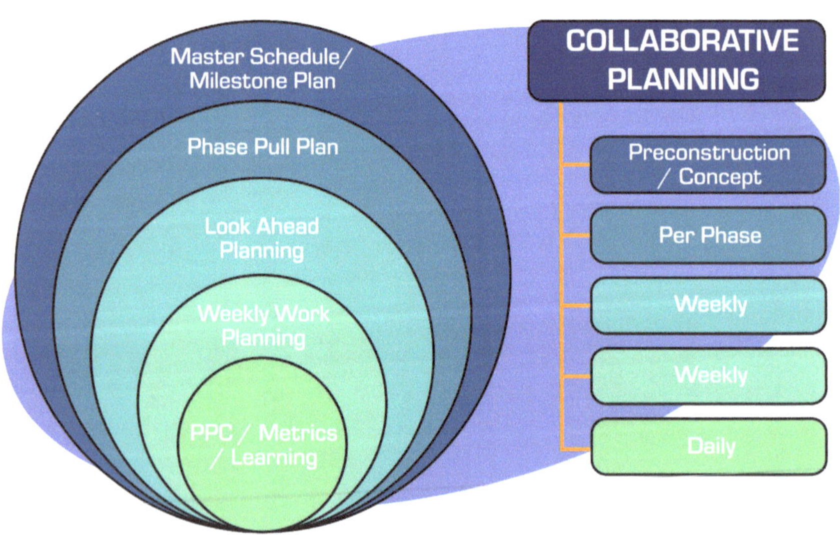

There are multiple platforms that can support LPS, both digital and physical. For the sake of this overview, we will assume we are not using a digital platform. It is also recommended that teams that are new to these tools first learn them using paper, markers, sticky notes, etc. before transitioning to a digital space, so that all team members understand the underlying purpose of each of the tools.

MILESTONE PLANNING/MASTER PLANNING

The Milestone, or Master, Plan is the foundation for all other planning to come. This may actually be completed without the input of the Last Planners, but it is becoming increasingly common to make this a more collaborative exercise, as well. It is not mandatory for this plan to contain excessive detail. The purpose of the Milestone/Master Plan is to confirm the feasibility of the overall schedule, identify long lead items for early action, and layout the project milestones for the Pull Planning process.

While developing this Master Plan, keep in mind that many of the assumptions, sequences, durations, and work breakdowns may (and very likely will) be changed throughout the Pull Planning process.

PHASE PULL PLANNING

Conventional wisdom tells us that all work should be performed as soon as it can be to speed up completion of a project. The principles of Lean Design and Construction challenge that assertion. Instead, we focus on handoffs of work from contributor to contributor (customers) based on trigger activities to provide exactly what is needed when it is needed. The concept of "working at the pull of the customer" provides many benefits to a project schedule, including a more logical work sequence, better workflow, and distribution of the workload over the entire value stream. Further, the detailed planning and coordination that go into the development of a Pull Plan reduce the likelihood of work being completed out of sequence, which often leads to many of our previously discussed wastes, such as defects, waiting, over-processing, and work-arounds.

Here is a step-by-step overview of the Phase Pull Plan process:

- Pull Plans are ideally developed 6 weeks before the beginning of the subject phase of work, and should cover approximately 10-14 weeks of work. By planning the work six weeks in advance, it will immediately become a part of the 6-Week Look-Ahead Planning process. 10-14 weeks is a manageable amount of work to cover in approximately four hours. (Experienced teams may need less time, while new teams may need some extra time.) Further, planning out too far into the future reduces the reliability of the assumptions made during planning.

- The planning session is coordinated by the leadership team on the project and needs to include all Last Planners that will have work during the phase. It is tempting to exclude Last Planners that don't have much work during the phase of work being planned, but resist the urge. Besides resulting in a well-developed execution plan, another benefit of pull planning is that all of the Last Planners understand and commit to the plan and their part therein, which results in better reliability and flexibility.

- The session opens with introductions and a group discussion of the start and finish milestones, most logical work breakdown (areas / batches) and flow of work, and general sequence of work. This "set-up conversation," which typically lasts 1-2 hours, ensures team understanding, gives Last Planners an opportunity to question and adjust assumptions that were made during the Master Planning, gets everyone using the same terminology

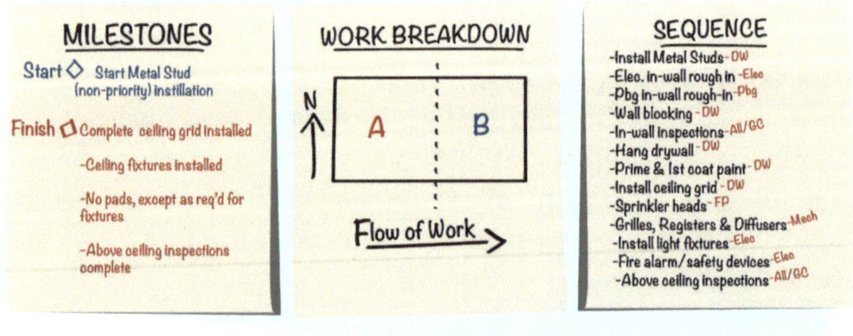

regarding areas of work, allows for conflict identification and resolution before beginning the detailed planning, and provides a guide for the Last Planners to identify their activities and triggers. It is often advantageous to have the set-up conversation a day or two before the rest of the pull plan.

• After the set-up conversation, each Last Planner is assigned a different color sticky note and prepares one for each activity.

Here is an example of a construction sticky note:

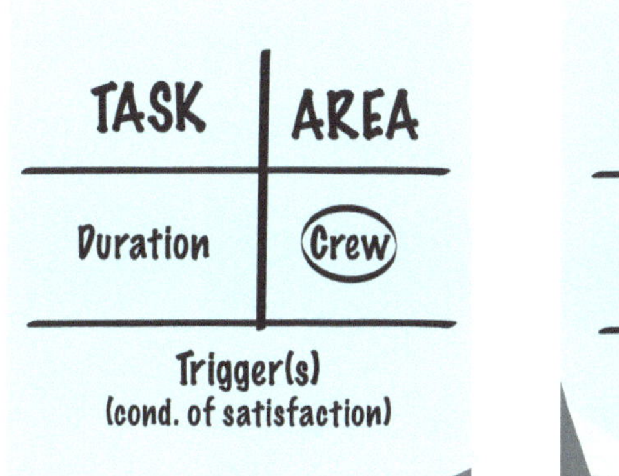

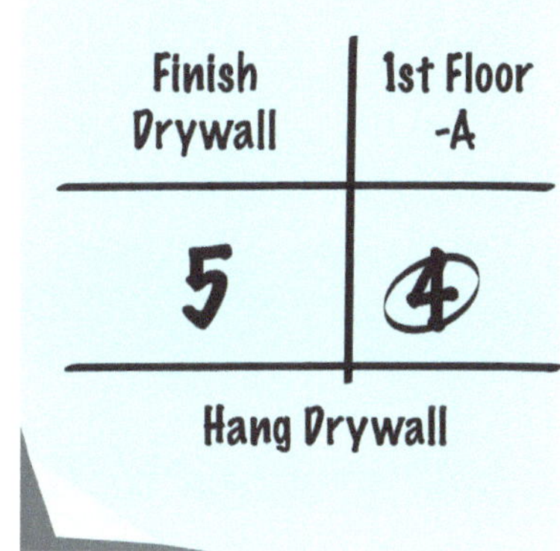

Here is an example of a design sticky note:

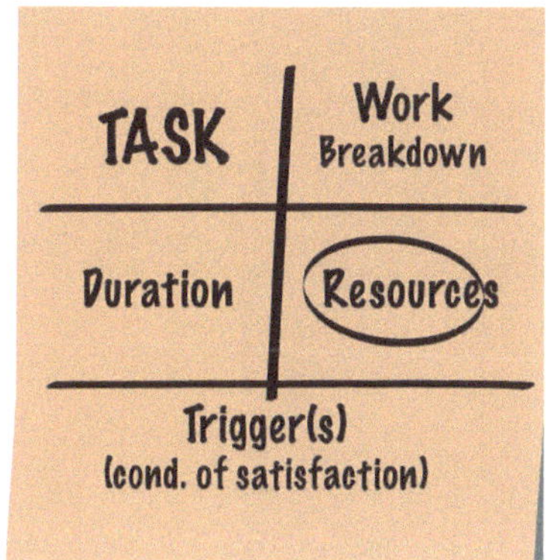

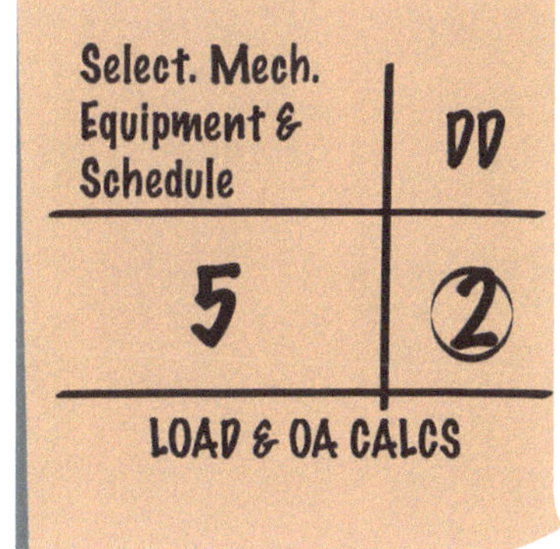

FILLING OUT YOUR STICKY NOTES

TASK: A description of the specific activity to be performed. In construction, this is pretty self-explanatory. "Hanging drywall" is a specific activity. It's a little tougher in design. It's easy to default to the deliverable being developed for the TASK, but we need to break those deliverables into the specific activities that need to be performed to complete the deliverable. When in doubt, make sure that your TASK contains an active verb (ex. Hang, install, rough-in, draw, write, field measure, survey, etc.).

AREA / WORK BREAKDOWN: Agreed-upon identifier for what area or portion of the work to which the task applies. Often different trades/disciplines will describe this differently (ex. Floor, shop drawing, pour area, deliverable, design phase, etc.). It is imperative that the group come to a consensus on how the work breakdown will be communicated.

DURATION: Number of days required to complete the task. The pull plan process is very visual and requires the ability to illustrate relative relationships between activities and show overlapping activities. Therefore, no sticky note should represent more than five days. If your activity requires more than five days, simply make multiple sticky notes. This may vary, based on specific project conditions, such as 6-day work weeks or day-by-day planning.

CREW / RESOURCES: The number of people required to complete the task within the identified duration. While including this information gives a general idea of how many people will be required on the project at a specific time, it is more important for the fact that it forces Last Planners to consider the resources that are required to complete the task in the committed time frame during the planning, hence adding to the reliability of the final plan.

TRIGGERS / CONDITIONS OF SATISFACTION: The trigger is the predecessor activity that releases the task to begin. This information is key, as it is the method by which we will work backward through the schedule to achieve the best sequence of work. Conditions of Satisfaction should be included if there is a specific requirement necessary for the work to begin (ex. All floors swept, dumpster on site, pre-task meeting).

• Once all Last Planners have made their sticky notes, the team reconvenes to "pull the plan."

Hang approximately 20' of plotter or butcher paper on the wall.

1. *Place a milestone marker at the far right of the paper for the end of the subject phase of work.*

2. *The Last Planner responsible for the final task places the sticky note(s) to the left of the milestone sticky and reads the activity information to the group.*

3. *The Last Planner responsible for the "trigger" activity places that sticky note to the left, and the process continues to the beginning of the phase.*

4. *When there are multiple tasks that need to be completed at the same time to progress, the task with the longest duration should be placed at the top. This will allow you to add up the top line to get the approximate critical path duration.*

5. *Once your team has placed sticky notes from the end milestone to the start milestone, ask if anyone is still holding any tasks. If so, confirm that they are not critical activities that were missed in the first pass, and then have the Last Planners place them where*

they belong in the sequence.

6. *That concludes the "first pass." Add up the duration of the top line – critical path – and compare it to the time available for the work to be done, according to the Master Plan.*

7. *As a group, walk through the plan from the beginning to the end, looking for logical problems for correction and opportunities for improvement (second pass). Again, add up the duration of the top line and compare it to the time available for work to be done.*

8. *Repeat passes until the team agrees that no further improvements can be made using the current assumptions.*

- If the phase duration is still not within the required duration, discuss the plan of action and next steps (ex. reconvene after additional information is gathered, consider additional crews or overtime, discuss the discrepancy with the owner, etc.). Record these next steps with assigned responsibility and promised completion dates.

- When the team agrees on the plan, enter it into the project scheduling software for distribution.

- Begin the Weekly Planning Cycle, with a focus on project execution according to the Pull Plan. (See Chapter 4 for a detailed example of a typical Weekly Planning Cycle.)

LOOK-AHEAD PLANNING / 6-WEEK LOOK-AHEAD PLAN (6WLAP)

Imagine, if you will, that our project is a ship moving through the ocean, and obstacles to our progress are icebergs. If we notice an iceberg when it's only 50 yards out, we're probably going to hit it. On our projects, hitting that metaphorical iceberg may lead to delays in work, added cost, work-arounds, rework, or safety incidents. But, what if we were "pinging" the water in our path far ahead of our current location? Then we would see the iceberg before it was an imminent threat and avoid hitting it with little to no impact on our estimated time of arrival. That is the idea behind Look-Ahead Planning.

Clearing constraints in a timely manner significantly improves the flow of work and reliability of the plans put in place. On a weekly basis (at the Weekly Work Plan meeting), the team considers the work that needs to be completed in the next six weeks to progress according to the Pull Plan. The onsite leader (General Superintendent, Design Lead, etc.) kicks off the conversation by describing the vision of where the project needs to be six weeks from now. The Last Planners should consider the work that they need to complete to get there and voice any concerns about obstacles (Constraints) to completing any of that work. Common constraints include:

OWNER DECISIONS TO BE MADE

REQUESTS FOR INFORMATION TO BE ANSWERED

PURCHASE ORDERS/CONTRACTS/WORK ORDERS TO BE EXECUTED

SPECIAL EQUIPMENT NEEDED

LABOR CONCERNS

PERMITS TO BE APPLIED FOR/RECEIVED

SUBMITTALS/SHOP DRAWINGS TO BE SUBMITTED AND/OR APPROVED

MATERIAL TO BE DELIVERED

CHANGE ORDERS TO BE APPROVED

BULLETINS TO BE ISSUED

INSPECTIONS TO BE SCHEDULED/CONDUCTED/PASSED

When a constraint is identified, it is added to the Constraint Log. The Constraint Log is, preferably, a white board in the LPS meeting space on which all of the constraints are recorded and tracked. It can also be kept in a spreadsheet, but remember that the idea is to keep these important issues in front of the team and easily accessible. If the team opts to only use a spreadsheet (ie. if the team is working remotely), they need to plan for making the Constraint Log, as well as the other LPS documentation, top of mind and accessible. For example, consider creating a project dashboard with all pertinent information at-a-glance. If the team maintains the Constraint Log in both white board and spreadsheet form, then a plan needs to be in place to ensure that they always match each other.

Here is an example of a typical Constraint Log:

Constraint Log

DESCRIPTION	Date Added	Date Needed	Champion	Date Promised	Date Cleared
Owner carpet selection for lobby	3/2	4/3	TS	4/1	
RFI #34 (AHU Alternative)	4/1	5/1	JB	4/21	
Curb cut permit for west entrance	4/1	4/16	TS	4/15	
Buy-out site irrigation	4/3	4/2	BR	4/21	
Final owner provided furniture list	3/27	4/15	TS	4/14	

Constraints may be turned RED, or otherwise distinguished, if they are within a week of being needed or already past the "Need By" date. This acts as a visual cue to the team that immediate action is required.

When a constraint is cleared, all affected Last Planners should be immediately notified. The constraint should remain on the Constraint Log until the entire team has been updated on the status during the Weekly Work Plan meeting.

See the Constraint Log Guide on the next page.

CONSTRAINT LOG GUIDE

DESCRIPTION: A description of the specific constraint to be resolved, as well as what work its resolution will release, if it is not well-known or obvious.

DATE ADDED: The date the constraint was added to the Constraint Log. Constraints can be added to the Constraint Log any time they are identified. It does not have to be during a Weekly Work Plan Meeting.

DATE NEEDED: The specific date that a constraint needs to be cleared in order to avoid disrupting the project plan. The date has nothing to do with whether it is a reasonable amount of time to get it cleared or what the project specifications state as expected RFI/Change Order/Submittal/Etc. times. The team should agree that, if the "Need By" date passes without the constraint being resolved, we will need to make adjustments to the plan to accommodate the issue. The goal is to identify constraints six weeks before they are a problem for the project progress, but that won't always be the case. If the team realizes that the project plan will need to be adjusted because the constraint wasn't cleared yesterday, then they need to put yesterday's date on the log. This should trigger the team to take further action.

CHAMPION: A member of the Leadership Team that is taking responsibility to get the constraint cleared. The Champion is not the individual or company that needs to take the action that clears the constraint (ex. Architect, Owner, Vendor, etc.). A member of the Leadership Team commits to "chasing down" a constraint until it is resolved to clear the way for work to progress. This is not an opportunity to "put the ball in someone else's court."

DATE PROMISED: The date by which the Champion commits to get the constraint cleared to allow work to progress. This should be a realistic date, and, therefore, may extend beyond the "Need By" date. If that is the case, the team should take immediate action to adjust the plan accordingly.

WEEKLY WORK PLANNING

The Weekly Work Plan (WWP) is the collaboratively developed day-by-day execution plan for the upcoming week. There are multiple strategies for developing this plan, but the key goals and requirements of the WWP remain the same:

- Based on the 6WLAP, Last Planners devise their individual plans for work activities that they will perform each day of the week.

  » Activities, as described, should be able to be completed by the end of the work week. This allows the team to accurately assess the reliability of the overall WWP and quickly recognize shortfalls. This often means breaking activities from the Pull Plan into smaller batches with specific extents (ex. Grid lines, room, drawings, etc.).

  » No activity should be included on the WWP that cannot be performed for any reason.

  » Last Planners should consider what work needs to be done to remain true to the Pull Plan and 6WLAP.

  » Last Planners should make note of any identified divergence from the Pull Plan, so the team can devise corrective measures.

  » Each Last Planner should strive to identify two or three Workable Backlog items each week.

    ◇ Workable backlog are activities that have no constraints, but do not need to be completed in the upcoming week to support progress and are therefore not a part of the WWP. These activities are essentially "Plan B," in the event that a planned activity cannot be performed or finishes early, freeing up the crew for additional work.

    ◇ Workable backlog needs to be identified during planning and discussed at the WWP Meeting to verify that moving to that unplanned activity will not adversely affect another trade/discipline in an unanticipated way.

- Last Planners submit their proposed WWP to the Leadership Team who then compile all of the WWP's into the preliminary

Team WWP.

- At the WWP Meeting, each Last Planner presents their plan to the entire team for comment and discussion. As adjustments are made, a member of the Leadership Team should note them on the team WWP. (If the team is using interactive WWP boards, the Last Planners make the adjustments themselves.)

- After the WWP Meeting, the Leadership Team distributes and posts the final Team WWP for the upcoming week.

- The WWP becomes the basis for discussion at the Daily Huddle/ Stand-up meeting.

As stated earlier, there are multiple ways to conduct Weekly Work Planning. The simplest and most common method for developing the WWP is with a spreadsheet, as seen here:

CORE TEAM & CONTRACTORS

Project Owner	Central Healthcare			
Project Name	Surgery Center			
Project Number	21-32			
Project Location	Cincinnati, Oh			
Project Lean Champion	Chris Berlage			

CATEGORIES OF VARIANCE

1	Safety Concern	8	Incorrect Duration
2	Information	9	Submittals/Approvals
3	Owner Decision	10	Equipment
4	Weather	11	Unforseen Condition
5	Prerequisite Work	12	Permit/Inspection
6	Labor	13	Coord / Plan Change
7	Materials	14	Other

WEEKLY WORK PLAN

TOTAL ACTIVITIES	
ACTIVITIES COMPLETED	
PERCENT PLAN COMPLETE	#DIV/0

WEEK OF Feb 21 Monday

ITEM #	LEVEL	AREA/WB	ACTIVITY DESCRIPTION	RESPONSIBLE Team/Company	RESPONSIBLE Person	COMMENTS, CONSTRAINTS, PREREQUISITES	MON FEB 21	TUE FEB 22	WED FEB 23	THU FEB 24	FRI FEB 25	SAT FEB 26	SUN FEB 27	DONE? YES	DONE? NO	REASONS FOR VARIANCE (Track in daily huddle)	Variance Category
1		Site	Selective Demolition of out buildings	GC	TC	Maint. Bldg empty by 2/21		5	5	5		X					
2		Site	Start clearing west side of site (complete 3/3)	GC	TC						3		X				
3		Pre	Selective Demolition of out buildings	Core	RG	Envelope cluster to present analysis at 2/24 Core Team mtg.				X							
4		Pre	Update meeting with Building Department	Core	MS				X								
5		Pre	Community progress update	Core	TM	Newsletter, Facebook, Twitter, Site posting					X						
6		Pre	Compile analysis of roofing design sets	ENV	RJ	Presentation of recommendation at 2/24 Core Team mtg.	X	X	X	X							
7		Pre	Compile furn. options for meeting with the Owner 2/25	INT	KD		X	X	X	X	X						
WORKABLE BACKLOG / ONGOING CLUSTER WORK FOR UPCOMING MILESTONES																	
		Site	Clear east side of site	GC	TC												
		Site	Begin removing spoils from the site	GC	TC												
		Pre	Continue working toward Envelope pkg · 3/15	ENV	RJ												
		Pre	Continue working toward Finishes pkg · 3/29	INT	KD												

The spreadsheet, widely used by organizations and projects implementing the Last Planner System, provides a template to identify activities to be performed and on which days they will be performed. There is also space for Workable Backlog, constraints/ comments, and Percent Plan Complete information, which will be described later in this chapter.

Another, more collaborative/hands-on, method is onsite WWP/6WLAP boards. To implement this method, we'll need to construct seven white boards with tape delineating the days of the week on each. To make this practical for the space available, we can add "sliders" at the top and bottom to allow the boards to move behind each other. For this to be an effective and efficient planning tool, the Leadership Team must populate the boards with the 6WLAP activities. During the WWP Meeting, each Last Planner stands at the board and places their activities for the upcoming week, making adjustments to coordinate with the other Last Planners' plans. Once the exercise is complete, and the Last Planners commit to the plan, it can either be left on the board for review throughout the week, it can be entered into a spreadsheet for easy access, and/or a picture can be taken for distribution and posting. *Be sure to include a space for Workable Backlog on the boards*. After the WWP Meeting, add the activities for the new sixth week on the last 6WLAP board.

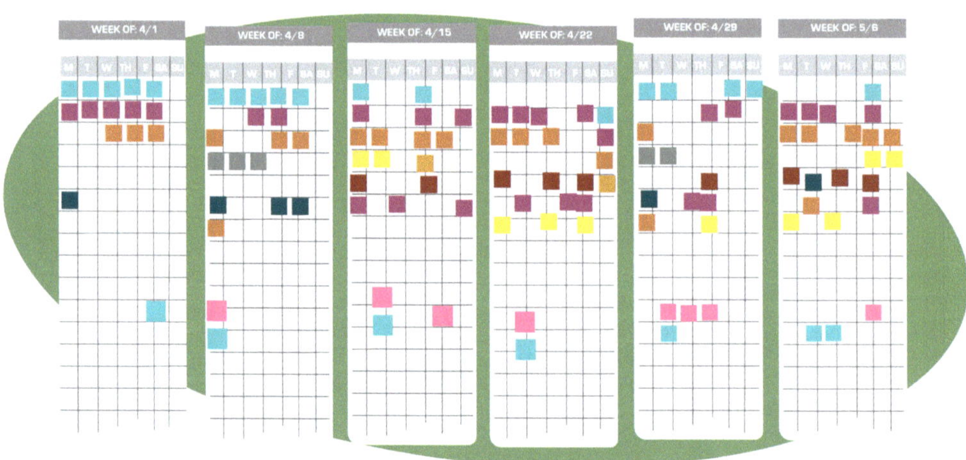

PERCENT PLAN COMPLETE (PPC) / VARIANCE TRACKING

Remember our key tenet of Continuous Improvement? Measuring PPC and tracking variance causes are key to achieving continuous improvement in the reliability of the commitments that team members make to each other.

Throughout the week, the team tracks which activities were completed according to the WWP and which ones weren't, with a simple YES or NO on the WWP sheet. Clearly, if an activity is completed later than committed, it would receive a NO. However, and this is very important, if an activity is completed earlier than committed, it also receives a NO. At the end of the week, the number of activities that received a YES is divided by the total number of activities included on the team's WWP to determine the percentage of activities completed reliably – the Percent Plan Complete.

CORE TEAM & CONTRACTORS

Project Owner	Central Healthcare
Project Name	Surgery Center
Project Number	21-32
Project Location	Cincinnati, Oh
Project Lean Champion	Chris Berlage

WEEK OF Feb 21 Monday

CATEGORIES OF VARIANCE

1 Safety Concern	8 Incorrect Duration
2 Information	9 Submittals/Approvals
3 Owner Decision	10 Equipment
4 Weather	11 Unforseen Condition
5 Prerequisite Work	12 Permit/Inspection
6 Labor	13 Coord / Plan Change
7 Materials	14 Other

WEEKLY WORK PLAN

TOTAL ACTIVITIES	7
ACTIVITIES COMPLETED AS PLANNED	5
PERCENT PLAN COMPLETE	71%

PPC ANALYSIS

ITEM #	LEVEL	AREA/WB	ACTIVITY DESCRIPTION	RESPONSIBLE Team/Company	RESPONSIBLE Person	COMMENTS, CONSTRAINTS, PREREQUISITES	MON FEB 21	TUE FEB 22	WED FEB 23	THU FEB 24	FRI FEB 25	SAT FEB 26	SUN FEB 27	DONE? YES	DONE? NO	REASONS FOR VARIANCE (Track in daily huddle)	Variance Category
1		Site	Selective Demolition of out buildings	GC	TC	Maint. Bldg empty by 2/21		5	5	5							
2		Site	Start clearing west side of site (complete 3/3)	GC	TC						3					Rain all day. Start Saturday & still finish 3/3.	4
3		Pre	Selective Demolition of out buildings	Core	RG	Envelope cluster to present analysis at 2/24 Core Team mtg.				X						Requested additional information for decision 3/4. Still on track for 3/15 ENV package.	2
4		Pre	Update meeting with Building Department	Core	MS				X								
5		Pre	Community progress update	Core	TM	Newsletter, Facebook, Twitter, Site posting					X						
6		Pre	Compile analysis of roofing design sets	ENV	RJ	Presentation of recommendation at 2/24 Core Team mtg.	X	X	X	X							
7		Pre	Compile furn. options for meeting with the Owner 2/25	INT	KD		X	X	X	X	X						
WORKABLE BACKLOG / ONGOING CLUSTER WORK FOR UPCOMING MILESTONES																	
		Site	Clear east side of site	GC	TC												
		Site	Begin removing spoils from the site	GC	TC												
		Pre	Continue working toward Envelope pkg - 3/15	ENV	RJ												
		Pre	Continue working toward Finishes pkg - 3/29	INT	KD												

To understand the answer to this question, the team first needs to see that a YES or NO is not inherently good or bad. It is merely a fact, and if that fact is that a task was not completed as planned, we, as a team, need to understand why so we can take action to be more reliable in the future. For this reason, when an activity is not completed according to plan, whether early or late, the follow-up should focus on identifying the root cause for the discrepancy. As a rule of thumb, this typically requires that the question, "Why?" be asked 5 times…

"**Why** was the activity completed late?"

"It took longer than usual."

"**Why?**"

"Because we didn't have the right equipment."

"**Why** not?"

"We didn't realize the extent of work that would be required."

"**Why** not?"

"We weren't able to do the necessary site investigation before we planned the work."

"**Why** not?"

"The Owner needed a week's notice, and we only gave them 3 days."

After this exchange, the team would identify the detailed reason, choose the category of "Coordination," and, if similar problems persist, actively seek opportunities to avoid the problem in the future.

But you're still asking yourself, "Why isn't completing an activity early a good thing?!?"

Despite the fact that we will diligently work to avoid positive and negative connotations for YES and NO, the fact remains that YES will be viewed as positive and NO will be viewed as negative. Therefore, by assigning a YES to work completed early, we are incentivizing the Last Planners to "sandbag," or add a day or two to each promised activity. When work is completed earlier than promised on the WWP, the Last Planner to whom the work is handed off is not likely to be able to adjust and start earlier than planned, so we will not gain anything when work is completed early.

We will, however, lose time each time work is completed late. That is why reliability is more important than beating our commitments. Our goal, as a team, is to continuously improve on the reliability of our promises. This means that sometimes activities will take longer than promised, and sometimes they will be completed sooner than promised. As long as we view each deviation as an opportunity to improve, the reliability of promises, and the PPC, will improve as the team becomes better at making reliable promises.

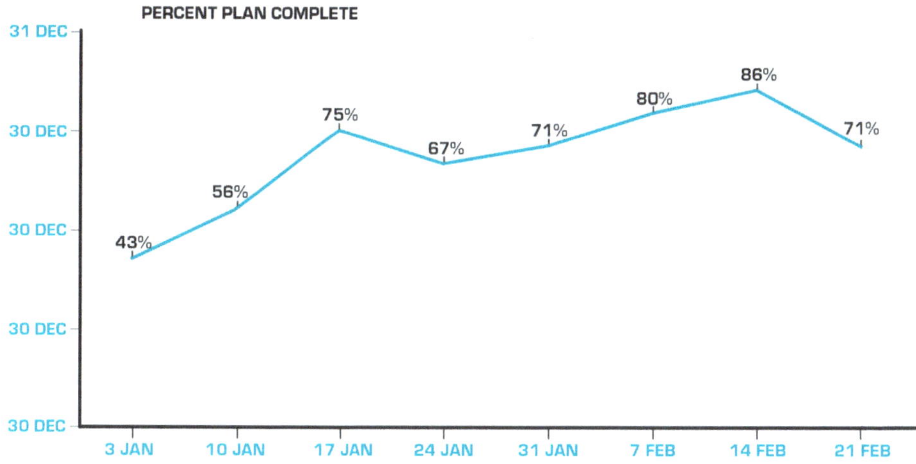

In the next chapter and Appendix A, you'll find explanations of the weekly and daily meetings that support collaborative use of these tools, as well as examples to help you get maximum value out of those meetings and the team.

4

"NO MATTER HOW GOOD YOU GET, YOU CAN ALWAYS GET BETTER, AND THAT'S THE EXCITING PART."

-Tiger Woods

CHAPTER FOUR

LAST PLANNER SYSTEM® OVERVIEW – MEETINGS

Chapter 3 gave an overview of the tools associated with the Last Planner System®. Deliberate and consistent employment of the LPS tools is key to a successful Lean project implementation.

The tools of the Last Planner System include:

MILESTONE PLANNING/MASTER PLANNING

PHASE PULL PLANNING

LOOK-AHEAD PLANNING

WEEKLY WORK PLANNING

PERCENT PLAN COMPLETE/VARIANCE TRACKING

The system is supported by two regular meetings (+ one prep meeting that improves the efficiency of the Weekly Work Plan Meeting):

WEEKLY WORK PLAN MEETING

DAILY HUDDLE/STAND-UP
LEADERSHIP TEAM WWP MEETING PREP

This chapter will focus on facilitation of the LPS meetings to optimize the benefits we reap from the tools we're using. First, let's look at a typical schedule of activities on an LPS project.

THE WEEKLY PLANNING CYCLE

Successful implementation of LPS requires that the project team develop a certain cadence of planning, communicating, and executing the work. Establishing a disciplined Weekly Planning Cycle is the key to this cadence. Here is an example:

MONDAY – Leadership team posts and distributes the 6-Week Look-Ahead Plan for the Last Planners to reference while completing their Weekly Work Plans.

TUESDAY – Leadership Team posts the Percent Plan Complete & Variance Tracker for the previous week.

WEDNESDAY BY NOON – Last Planners submit their completed Weekly Work Plan for the upcoming week to the Leadership Team.

WEDNESDAY AFTERNOON – Leadership team compiles all Last Planner WWP's into a Team WWP.

WEDNESDAY AFTERNOON – Leadership team meets (approximately 30 minutes) to prepare for the Weekly Work Plan Meeting.

THURSDAY – Leadership team and Last Planners convene for Weekly Work Plan Meeting and finalize the Weekly Work Plan for the upcoming week.

FRIDAY – Leadership team distributes and posts the Weekly Work Plan for the upcoming week.

DAILY AT 2PM – Last Planners attend the Daily Huddle / Stand-up meeting on site.

WEEKLY WORK PLAN MEETING

The Weekly Work Plan Meeting (WWP Meeting) is the culmination of the team's weekly planning efforts. It is the team's opportunity to share a weekly update on cleared and outstanding constraints, identify new constraints, review past performance, and plan next week's work. It is also the time during which the team needs to ensure that the short- and medium-term planning aligns with the Master Plan and Phase Pull Plans.

Almost every construction project on the planet holds a weekly project meeting. The Weekly Work Plan Meeting is probably different than the meetings to which you are accustomed. Often weekly project meeting agendas include in-depth discussions of all current and past issues on the job, review of the RFI log, submittal log, and bulletin/change order log, and any number of other updates, conversations, and admonitions that the Leadership Team wants to convey to the project team. They also tend to "last as long as they last."

As a sign of RESPECT (see the Six Tenets of Lean Construction), the WWP Meeting – and any other meeting held on a Lean project – starts on time and ends on time. The vast majority of projects need only one hour to cover the required topics of a WWP Meeting. It is the rare large, complex project that requires more time. If you find that your WWP Meetings are lasting more than 60 minutes, verify that your team is appropriately preparing for an efficient meeting and that you are not trying to address topics outside of the prescribed agenda. If after making adjustments, you still feel that your team needs more time, change the stated times to 90 minutes, and continue to start on time and end on time.

You may have noticed that the sample Weekly Planning Cycle above has the WWP Meeting scheduled for Thursdays. It is often recommended to hold this meeting on Wednesday afternoon or sometime on Thursday. The reason for that recommendation is that, by that time in the week, the Last Planners should have a relatively good idea of what they will complete in the current week, and will also still have enough time to make adjustments (order material, request needed labor and equipment, etc.) before the upcoming week begins. This timing can be adjusted to accommodate the needs of the project and team. Also, keep in mind that, although Monday through Sunday is typical, your WWP can span any seven days (ie. Wednesday through Tuesday). As always, the keys are planning, communication, and consistency.

Every meeting should have a stated purpose and agenda with allotted times for each discussion. This takes planning, but will help to ensure an efficient and effective meeting.

As stated above, the purpose of the WWP Meeting is to update the team on cleared and outstanding constraints, identify new constraints, review past performance, and plan next week's work. Having a clear understanding of the purpose makes it much easier to table discussions outside the purview of the agenda for follow-up when they arise. Teams should use a "Parking Lot" to ensure that follow-up occurs. The Parking Lot is ideally a white board on which the team records discussion topics and a Leadership Team member champion to make sure the follow-up occurs. To be clear, the Parking Lot is for follow-up conversations/meetings/etc., whereas the Constraint Log is to record specific actions that need to take place to allow work to progress according to the plan.

So, what does the agenda for a WWP Meeting look like? It is so simple and so consistent that the Leadership Team doesn't even need to print it out. Post it on the wall, and facilitate the same general discussion each week. Here is a sample WWP Meeting agenda:

WWP MEETING AGENDA EXAMPLE

(5 MIN.) GENERAL PROJECT INFORMATION – SAFETY/SECURITY/OPERATIONS

(3 MIN.) UPDATE ACTIONS TAKEN ON LAST WEEK'S PLUS/DELTA (+/Δ)

(5 MIN.) REVIEW LAST WEEK'S PERFORMANCE

- » Team PPC
- » Three biggest variances
- » Trends in variances – What are we doing about it?

(10 MIN.) LOOK-AHEAD PLANNING

- » Report on Updates to Constraint Log / Parking Lot
- » 6WLAP / Where should we be in six weeks?
- » What could stop us from getting there? / What constraints should we add to the Constraint Log?

(10-25 MIN.) PRESENT & FINALIZE TEAM WWP

(15 MIN.) ROUND ROBIN

- » New/Other issues
- » +/Δ

"MEETING ADJORNED"

PLUS/DELTA (+/Δ) DEFINITION

A continuous improvement discussion preformed at the end of a meeting, project, or event and used to evaluate the session or activity. Two questions are asked and discussed.

Plus: What produced value during the session?

Delta: What could we change to improve the process or outcome?

Definition from the Lean Construction Institute Glossary

Different members of the team fill different roles during the meeting:

FACILITATOR: A member of the Leadership Team that keeps the meeting on track according to the agenda. This can be anyone that has the skills to keep the meeting moving in the right direction.

LEADER: The member of the Leadership Team that will present the 6-Week Look-Ahead "vision" and play a key role in the WWP coordination discussion. This is often the General Superintendent or Design Lead for the project.

CONSTRAINT/PARKING LOT LEAD: The member of the Leadership Team that takes responsibility to capture Constraints and Parking Lot items throughout the meeting. This should be someone that can identify Constraints and Parking Lot items when they come up, and is comfortable pausing the meeting to record the required information.

WWP RECORDER: The member of the Leadership Team that makes the required adjustments to the WWP during the WWP discussion.

LAST PLANNERS: The trade foremen/discipline leads that actively participate in the discussions, present their proposed WWP's to the team, and collaborate to develop the optimal Team WWP for the upcoming week.

*Note: The same individual can fill multiple roles, but the assignments need to be determined before the meeting starts. (See PREP MEETING for more details.)

Once the meeting begins, the goal is for the Last Planners to actively collaborate with one another. The Leadership Team's role is to guide and encourage that collaboration, and to keep the team focused on creating a short-term plan that supports the Master Plan and the Phase Pull Plans. (See Appendix A for Sample Meeting Discussions.)

DAILY HUDDLE / STAND-UP MEETING

Up to this point, we've talked about a couple of concepts and tools that are key to achieving FLOW (one of the Six Tenets of Lean) in our work. First, while developing the Phase Pull Plans and Weekly

Work Plans, the team focuses on hand-offs from one trade/discipline to the next, completing work at the right time to release the next activity. Second, the team utilizes Look-Ahead Planning and the Constraint Log to identify and clear obstacles to progress before they impact the plan. The Daily Huddle Meeting further supports project workflow by facilitating daily conversation between the Last Planners, allowing for real-time updates to the WWP and Constraint Log, and providing an opportunity for Last Planners to communicate readiness for handoffs, variances to the plan, and needs from other team members.

The Daily Huddle typically lasts 15 to 20 minutes and takes place at the site of the work, assuming the participants are all on site. Adjustments to the location may be made to accommodate the team make-up, but the length of the meeting should remain consistent... and short. It is typically facilitated/led by the General Superintendent or Lead Designer.

The sample Weekly Planning Cycle above shows the Daily Huddle taking place each day at 2PM. The recommendation to hold this meeting in the early afternoon is similar to the recommendation to hold WWP Meetings on Wednesday or Thursday. The reason is that, by that time in the day, the Last Planners should have a relatively good idea of what they will complete by the end of the day, and will also have enough time to make adjustments before the next day. This timing can be adjusted to accommodate the needs of the project and team. Many teams prefer to hold daily huddles first thing in the morning. By shifting to the afternoon, everyone can hit the ground running at the beginning of the shift, the Last Planners are already aware of any changes to the plan, and they aren't being pulled away from their crews/teams at that crucial time in the day.

Again, every meeting should have a stated purpose and agenda. As stated above, the purpose of the Daily Huddle is to facilitate daily conversation between the Last Planners, allow for real-time updates to the WWP and Constraint Log, and provide an opportunity for Last Planners to communicate readiness for handoffs, variances to the plan, and needs from other team members.

The agenda for the Daily Huddle is less prescriptive than the WWP Meeting, but it is still the facilitator/leader's responsibility to ensure that discussion remains relevant to the purpose and the meeting is complete within 20 minutes. Utilize the Parking Lot as required to achieve these goals. The meeting opens with a quick update of constraints that have been cleared since the previous day, and then becomes a "round robin" among the Last Planners to report what work will be ready for handoff at the end of the day, anything that was not / will not be completed according to the WWP, and requests for assistance / newly identified constraints. (See Appendix A for Sample Meeting Discussions.)

LEADERSHIP TEAM WWP MEETING PREP

The WWP Meeting and the Daily Huddle are integral parts of the Last Planner System. This prep meeting is the Leadership Team's method to ensure that the WWP Meeting is as meaningful, effective, and efficient as possible. By taking the time to plan and prepare for the WWP Meeting, the Leadership Team is showing RESPECT for the rest of the team and helping to enhance project outcomes.

The prep meeting takes place prior to the WWP Meeting and typically lasts about 30 minutes. The attendees include the members of the Leadership Team that will attend the WWP Meeting, as well as any other Leadership Team members that should have input in the outcome of the meeting. The topics that should be covered include:

- Who will facilitate the WWP Meeting?

- Who will be the leader of the WWP and Look-Ahead discussions?

  » Typically the General Superintendent / Lead Designer

- Who will capture the Constraints and Parking Lot items on the respective logs?

  » This needs to be someone that can recognize a constraint and feels comfortable stopping a conversation that should be added to the Parking Lot.

- Prepare for the +/Δ update

  » What was done about last week's identified +/Δ?

- Prepare for the Constraint Log / Parking Lot update

  » Know the status of all items, and be prepared to give a quick report that includes:

    ◇ Items cleared / resolved since last week;

    ◇ Items added since last week; and

    ◇ Items that are approaching or already past the need-by date and the plan to address them.

- Compile comments and questions for the WWP discussion

  » Leadership Team members should review the WWP prior to the prep meeting.

  » Leader should prepare notes to guide the WWP discussion based on their observations and input from the rest of the Leadership Team.

    ◇ TIP: It may be helpful to put questions/comments on the compiled WWP in a different font to ensure they are addressed during the meeting.

The described LPS meetings are key components to achieving collaboration and overall project success. See Appendix A for sample scripts that will help the project team optimize the value and efficiency of these meetings.

5

"GREAT THINGS ARE DONE BY A SERIES OF SMALL THINGS BROUGHT TOGETHER."

-Vincent Van Gogh

CHAPTER FIVE

TARGET VALUE DELIVERY OVERVIEW

Not every project employs Target Value Delivery (TVD), and not every project is organized to exploit all of the benefits of TVD. However, every project can realize benefits from using SOME of the tools of TVD. For that reason, we're going to provide an overview to establish a general understanding of the concepts.

It is typically accepted that as value is increased on a project, so is the cost. TVD aims to turn that assumption on its head. A committed TVD team focuses their efforts on reducing the overall cost of the project, while also increasing the value provided to the customer. We achieve this by collaboratively setting aggressive targets for budget, schedule, safety, and quality, and then making those targets, as well as the owner's Value Statement and the team's Conditions of Satisfaction, inputs to the design decisions made for the project. No design decision is finalized until it has been weighed against the project targets.

On a traditional project, design is completed in defined phases after which the team engages in an estimating exercise to assess whether or not the current design fits within the prescribed budget. If the answer is no, as it often is, the team then performs "value engineering" to reduce the cost. Value engineering usually requires a reduction in the value being provided. It is also rework on the part of the designers, which, as we discussed in Chapter 2, is pure waste. In TVD, since no decision is made without first being compared to the target cost, the final design fits into the project budget and no value engineering is required. This is achieved through a combination of set-based design and continuous conceptual estimating.

> ## TARGET VALUE DELIVERY DEFINITION
>
> **Target Value Delivery**: A disciplined management practice to be used throughout the project to assure that the facility meets the operational needs and values of the users, is delivered within the allowable budget, and promotes innovation throughout the process to increase value and eliminate waste (time, money, human effort).
>
> *Definition from the Lean Construction Institute Glossary*

TVD PHASES

Like many traditional jobs, a project utilizing Target Value Delivery is broken into four distinct phases (Business Case, Validation, Value Delivery, Value Post Construction). The difference is in the fact that the component activities of a TVD project overlap and span multiple phases, as different parts of the final deliverable are selected, designed, and constructed at different times throughout the process.

PHASE 1 – BUSINESS CASE

The Business Case is, in essence, a question posed by the Owner to the TVD team regarding the feasibility of the project. It becomes the subject of the Validation phase. For example:

"Can we build a 200 bed clinical hospital in Chicago, IL for $185,000,000 in 3 years?"

"Can this casino renovation generate 25% increased monthly income within 2 years for $15,000,000?"

"Can we build a homeless shelter to serve 17 families in Jacksonville, FL for $6,000,000 while utilizing local contractors and trades people?"

The Business Case is developed by the project Owner, perhaps with assistance from the Construction Manager and/or Project Architect, and is provided to the project team for validation. The two documents that must be provided are:

- **Allowable Cost** – the absolute maximum that can be spent on the entire project, including Owner costs; and

- **Value Statement** – definition of the Owner's desired outcomes from the project – high level statements of the required value to be achieved, developed with the input of all project stakeholders. Value requirements should not be ranked or weighted, as they are ALL equally important and will guide the target setting and decisions throughout the TVD process.

PHASE 2 – VALIDATION

The fact that validation is a key component in itself distinguishes TVD from traditional project delivery. When a traditional project team is formed, there is an assumption that the project will go forward, which often leads to value reduction to achieve the required budget. With TVD, the team spends time and resources to first assess their ability to deliver the required value within the Allowable Cost. If the answer is no, the project does not proceed. The Owner then has the option to cancel the project or make adjustments to the Business Case and begin the Validation process again.

If the project is deemed feasible, the required outputs from the Validation phase include:

- **Scope Definition** – The scope of the project evolves throughout the Validation Phase, beginning with development of the initial cost model (see next page), during which the team will identify initial assumptions about size, arrangement, and materials, but should not include final design decisions. Final decisions will be made during set-based design. Traditional design (drawings and specifications) takes place during the Value Delivery Phase.

- **Conditions of Satisfaction (CoS)** – Based partially on the Value Statement received as part of the Business Case, the CoS are a list of required outcomes to consider the project a success for all team members, including the Owner, CM, Designers, and Trade Partners. Creation of the CoS is a collaborative exercise that includes input from the entire team.

  The CoS will guide decisions throughout design and construction, as well as provide a basis for periodically measuring team performance and making adjustments as needed. Conditions of Satisfaction need to be measurable (Qualitative or Quantitative) and are not ranked, as they are all equally important to the success of the project.

  Once the team has agreed on the Conditions of Satisfaction, they must devise a method for measuring performance in each area. This could be simply running a report for some or sending out a survey to the team members for others. Finally, the team should schedule regular intervals to assess team performance against the CoS and commit to taking quick action to resolve any items that are outside of the acceptable threshold for success.

CONDITIONS OF SATISFACTION TIP

Conditions of Satisfaction should be *outcome-based*, not *solution-based*. For example, the condition should be, "Zero recordable accidents," as opposed to, "We will perform safety audits on a weekly basis."

While the safety audits may be part of the route to achieving the goal, the outcome is the actual required condition. Stating required outcomes gives the team the flexibility to devise solutions and adjust, as required, throughout the project.

- **Target Cost** – The overall cost that the team agrees to pursue for the project, the Target Cost is set below the best-case-scenario cost of a traditional project, based on benchmarking performed during the Validation phase. Since decisions are made only after being compared to the Target Cost structure, an aggressive

POTENTIAL CONDITIONS OF SATISFACTION

OWNER VALUES:
- "We will utilize 25% local workforce for construction." (Quantitative)
- "We will maintain a good relationship with our neighbors during construction." (Qualitative)

SCHEDULE:
- "We will complete the envelope, roof, and windows before September 30th to reduce temporary measures and cost for construction heating." (Quantitative)
- "We will complete the construction, punchlist, and all closeout activities by February 27th." (Quantitative)

SAFETY:
- We will achieve accident and incident rates below the national averages." (Quantitative)
- "Our craft force will incur zero 'significant' injuries." (Quantitative)

QUALITY:
- "We will devise a strategy to "error-proof" enclosure of mechanical spaces, such as walls, ceilings, and shafts, leading to zero re-work due to missed components and/or inspections." (Qualitative/Quantitative)
- "This project will have no more than 100 RFI's (Requests for Information) due to design errors." (Quantitative)

TEAM HEALTH:
- "We will maintain high team morale throughout the project and take corrective action if the average team member satisfaction rating falls below 4 on a scale of 1-5." (Qualitative)
- "Meeting quantity, timing, content, and required attendees will be assessed regularly and adjusted as needed to ensure a respectful and meaningful use of team member time." (Qualitative)

TEAM MEMBER SUCCESS:
- "Every team member will earn a fair profit." (Quantitative)
- "This team will be asked to work together on a future project." (Qualitative)

Target Cost naturally drives the team to actively seek innovation at every step along the way.

It is important to note that the Target Cost is set collaboratively by the project team during the Validation phase. It is not determined by the Owner during the Business Case.

During the Validation Phase, the project team performs multiple activities, the outputs of which serve as the foundation for further project development in future phases:

FORM THE PROJECT TEAM

As we discussed earlier, TVD principles and tools can be used on any project with any contract type, but as the collaborative nature of the agreement increases, so do the opportunities to use more components and optimize the potential benefits of TVD. Despite the contract type being used, the earlier the project team is brought on the better. The best illustration of this fact is the MacLeamy Curve, shown here:

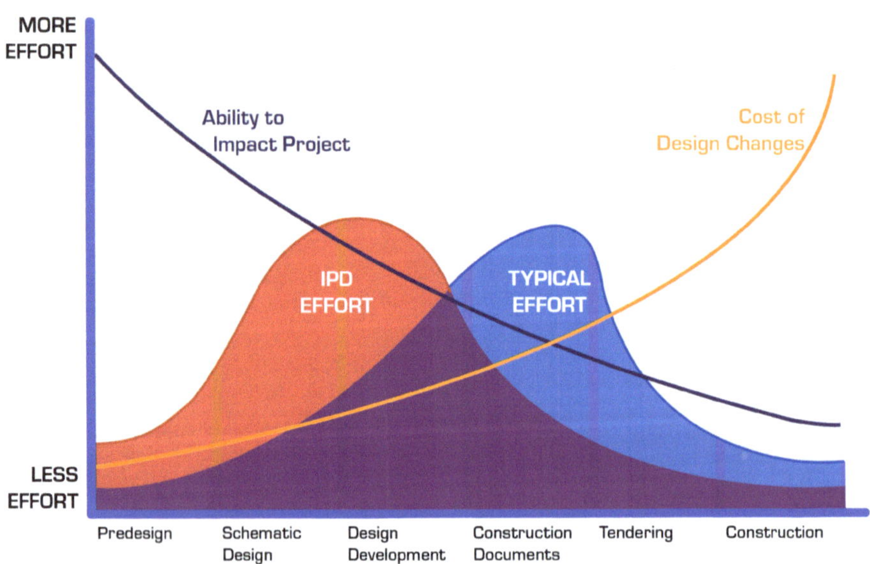

Used with permission from Patrick MacLeamy, FAIA

On a traditional project, the majority of the team is engaged after a bulk of the design is already complete, a point in the project during which any change will require a significant expenditure of money and resources. When the team (Owner, CM, Designers, Trade Partners) is formed before design occurs, they are able to provide invaluable input at a time when changes and corrections are relatively easy to make.

So, how do we select our team members? Ideally, and if the prescribed procurement method allows for it, team member selection will be based on:

- **TECHNICAL QUALIFICATIONS;**
- **EXPERIENCE WITH / WILLINGNESS TO PARTICIPATE IN A PROJECT EMPLOYING LEAN AND TVD TECHNIQUES;**
- **EXPERIENCE WITH SIMILAR PROJECTS;**
- **PROPOSED INDIVIDUALS' FIT WITH THE ESTABLISHED TEAM CULTURE; AND**
- **LASTLY, PROPOSED COSTS.**

There are multiple methods that can be used to select the appropriate team members, including:

- **Owner Selection** – A common selection method, after proposals are received and reviewed, potential team members are shortlisted and interviewed by the Owner for addition to the team.

- **Previously Successful Team** – The Owner elects to engage an existing team from a previously-successful project. (The dream!)

- **Progressive Team Building** – In this method, once a team member is selected, they become a member of the selection team for each additional team member. A major benefit of this method is that the team culture is being further developed with each new selection, and that culture becomes a key component for future selections. It is unlikely that a member will be added that doesn't "fit" with the rest of the team.

ORGANIZE THE TEAM/RAPIDLY CREATE A HIGH PERFORMING TEAM

The project team is the engine that drives any project. A high performing team encompasses the six tenets of Lean:

- **RESPECT FOR PEOPLE;**
- **OPTIMIZE THE WHOLE;**
- **GENERATE VALUE;**
- **ELIMINATE WASTE;**
- **FOCUS ON FLOW; AND**
- **CONTINUOUS IMPROVEMENT**

Further, a high performing team is aligned to break down traditional barriers to collaboration and enthusiastically drive toward a common goal. This phenomenon does not "just happen" because you have selected quality team members. A truly high performing team is the result of deliberate, planned organization and activities, and a sustained mutual effort to keep the team members aligned and engaged.

This team dynamic ideally begins with a multi-day project kick-off session with all members of the project team in attendance. Depending on team availability, it can also take place at separate times, but it is important that it is done as early as possible, once the team members are selected. The agenda should be structured to include designated time for topics such as:

- **Introductions** – Create an opportunity for the team members to get to know each other as individuals.

- **Project Overview / Review of the Business Case** – This presentation should not only address the business side of the project requirements, but also the meaning behind the project. This is the first opportunity to begin aligning the team behind a common mission.

- **Relevant Training** – Set the team up for success by providing the training necessary for them to perform effectively together.

- **Personality Assessment** – Many teams choose to conduct personality assessments for team members, and then review the results at the project kick-off. Sharing this information helps participants understand each other and the best ways to communicate.

- **Conditions of Satisfaction** – This facilitated, collaborative session establishes the team's CoS, providing the foundation for decision-making before work begins.

- **Big Room Methodology** – Take time to deliberately create your team's method for collaboration. Your Big Room approach may mean co-location, remote meetings, or a hybrid of the two, but it requires planning, consistency, and regular re-evaluation. Your Big Room plan will set the rhythm for the project throughout design and construction.

- **Big Ideas Session** – Give the team an opportunity to work together to brainstorm ways to add value and/or save costs, as well as identify potential risks to the project. Then assess the value and likelihood of the risks and opportunities to target the highest value options to pursue. Big Ideas sessions should take place multiple times throughout the project timeline, beginning with kick-off.

- **Collaborative Design Session** – This is the first step to aligning your entire team (Owner, CM, Designers, and Trade Partners) to develop set-based design options for the project.

- **Initial Pull Plan** – Since you'll likely be presenting LPS® training during the project kick-off, this is the ideal time to hold the first pull plan session – while the purpose and procedures are still fresh in everyone's mind. Additionally, since deadlines will be looming, it is the appropriate time to begin planning for the first phase of work.

- **Initial Work Cluster Formation** – TVD operates on a system of trust, respect, collaboration, and distributed leadership.

  » <u>The Executive Team</u> is made up of senior-level members from each participating organization, and are responsible for oversight and input on issues that cannot be resolved in the Core Team.

  » <u>The Core Team</u> consists of management-level members from each participating organization, and are tasked with the day-to-day operations. Project decisions are made by the Core Team, unless they must be elevated to the Executive Team.

» <u>Work Clusters</u> are multi-disciplinary teams, each responsible for a specific part of the project. A Work Cluster may focus on a specific system, such as HVAC or building envelope, or a specific project challenge, such as Safety or Prefabrication. These teams gather and analyze information on multiple options and make recommendations to the Core Team for final decision-making. Work Clusters should be formed and dissolved according to the needs of the project.

During kick-off, team members should identify the initial Work Clusters required and their members. Note that individual team members may play a role in multiple Work Clusters.

- **Next Steps Discussion** – Take the time to summarize the action items identified during kick-off and assign (via volunteer) responsibility and deadlines.

- **Plus / Delta (+ / Δ)** – Ask for feedback from the attendees to help improve future meetings and the overall project.

- **Team Building** – Be sure to set time aside, during the day and/or after hours, for team members to spend time getting to know each other as individuals.

- **Establish the Initial Cost Model** – The project Cost Model is the high level budget that will be used to track estimated and actual costs, guide decision making, and signal that there are risks and/or opportunities for the team to assess throughout the life of the project. In TVD, the Cost Model is a living document that is updated in real time as information becomes available.

COST MODEL DEFINITION

Cost Modeling - Developing a model of the cost components and systems specific to a project and structuring it in a manner that the components and system costs can be continually updated either via benchmarks, metrics, or detailed estimates to provide the team with a constantly up-to-date cost model for the project. In the TVD environment, the cost model should allow for projecting 'what-if' scenarios based on value decisions that have yet to be made.

Definition from the Lean Construction Institute Glossary

Project Cost Model

BUDGET CATEGORY (BUCKET	TARGET COST	CURRENT ESTIMATE	VARIANCE
Owner and Costs	$220,000	$220,000	$0
Owner Admin Costs	$52,000	$60,000	$8,000
Owner FF&E Costs	$102,000	$100,000	($2,000)
Permitting	$10,000	$9,990	($10)
TOTAL OWNER COSTS	$384,000	$389,990	$5990
General Conditions	$20,000	$185,000	($15,000)
Site	$512,000	$509,200	($2,278)
Foundations	$98,000	$94,635	($3,365)
Structure	$87,000	$96,620	($9,620)
Building Envelope	$275,000	$322,856	$47,856
Mechanical	$2,100,000	$1,994,780	($105,220)
Electrical	$950,000	$964,000	$14,000
Plumbing	$415,000	$402,900	($12,000)
Fire Protection	$132,000	$135,500	$3,500
Finishes	$740,000	$743,000	$3,000
Accessories	$76,000	$76,000	$0
TOTAL DESIGN & CONSTRUCTION COSTS	$5,585,500	$5,525,013	($59,987)
GRAND TOTAL	$5,969,000	$5,915,003	($53,997)

Risk and Opportunity REGISTER

ITEM DESCRIPTION	RISK	OPPORTUNITY	BUCKET	VALUE/CoS	%
Steel escalation	$20,000	- - -	Structure	N/A	35%
Smart Board Classrooms	- - -	$30,000	Accesories	Environment of Learning	50%
Rebate from roof manufacturor	- - -	($15,000)	Envelope	N/A	90%
Sponsor for Football Field	- - -	($100,000)	Site	Community Involvement	75%
Unforeseen underground vault at west site	$150,000	- - -	Site	N/A	65%
Pool on Roof	—	$65,000	Envelope	Non Identified	0%

The initial Cost Model is developed in the Validation phase, before any design elements have been selected. Therefore, the project team must gather Benchmark information from existing projects, normalize that information to match the conditions of the subject project, and identify risks. This exercise is not to be taken lightly and requires the input of all team members. The initial Cost Model is the main basis for the team's decision of whether the project is a GO or a NO-GO.

PHASE 3 – VALUE DELIVERY

Once the team determines that the project can go forward, the primary goal is to advance the work. Using the Last Planner System® to collaboratively plan and track activities and handoffs between design disciplines and trades, the team develops a final design and constructs the project.

The processes and activities initiated during Validation continue throughout the life of the project. The initial Cost Model and scope are the foundation for the work done during Value Delivery, while the Target Cost and Conditions of Satisfaction are considered as inputs to each decision made.

With traditional delivery methods, design progresses through defined phases at the end of which a comprehensive estimating exercise is performed to assess alignment with the project budget. The outcome is often a design that exceeds the budget, and the team must begin value engineering, or reducing/minimizing the scope to meet the intended budget. This always results in waste in the form of rework by the designers and often results in a reduction of value delivered.

Conversely, in TVD, no design decision is made until it has been compared to multiple options and weighed against the requirements of the Target Cost and Conditions of Satisfaction. Therefore, the final design adheres to the cost and value requirements of the project, eliminating the need for design rework.

What is described above requires skills that can be learned and nurtured, but should be recognized as a clear departure from traditional methods. Team members need to be trained, coached, and supported in the implementation of these new methods:

- **Set-based Design** – Multiple options for each design element are identified and analyzed by the appropriate Work Cluster. The analysis includes a comparison of key characteristics against the Conditions of Satisfaction and Target Cost. Final design decisions are made by the Core Team and incorporated into the project documents.

- **Conceptual Design** – For each option to be analyzed by the Work Clusters, the designers must provide a description of the general requirements for the subject system, including rough sketches, if necessary.

- **Continuous Estimating** – Unlike the traditional approach of receiving design documents, performing quantity take-off, and assigning costs, estimating resources on a TVD project need to hone their ability to visualize, communicate, and assign costs to general scope descriptions of multiple options. Further, this process must be completed quickly to support the set-based design and advancement of work.

Trust and respect are key. On traditional projects, while multiple sources may provide cost input, the verification and compilation of estimating efforts commonly falls on one or a small group of individual estimators. The Core Team must be able to trust the information developed by the Work Clusters. Otherwise, the project will experience a bottleneck in the estimating process.

Both design and construction are planned and executed using the Last Planner System and form an integrated project process. Design decisions are made to meet the needs of construction, and the construction team members participate in the design process as members of Work Clusters, ensuring the means and methods are considered.

During the Value Delivery phase, project components are selected and integrated into the design, as well as constructed, at the "last responsible moment." This means that the team must fight their natural instinct to complete and finalize work "as soon as possible," which often leads to decision-making without all of the pertinent information and analysis, and work being performed out of sequence. This will almost always lead to integration of design elements that don't adhere to the Target Cost or CoS, errors, rework, and variation in the workflow. Using the phase Pull Plans, the team can identify the best time to perform activities to advance the work reliably.

PHASE 4 – VALUE POST CONSTRUCTION

The true value of a project isn't realized until it is complete and in operation. On traditional projects, the project team often begins to disperse before the Owner takes occupancy, only to return if there are problems. In TVD, it is incumbent on the project team members to commit to regular, scheduled follow-ups to identify and correct failures in the value delivery. This requires a comparison of actual outcomes to the Target Value and Conditions of Satisfaction.

Further, project completion is the time for a final retrospective to capture team member input on what went well and what could be improved on future projects. This is one last chance to pursue ***continuous improvement*** with this team.

6

"COMING TOGETHER IS A BEGINNING, STAYING TOGETHER IS PROGRESS, AND WORKING TOGETHER IS SUCCESS."

-Henry Ford

CHAPTER SIX

ONBOARDING & TEAM BUILDING

In Chapter 5, we talked about team formation and team building. While all of the tools discussed in the overview of Target Value Delivery can be applied as part of any project delivery model, we're going to spend some additional time on these topics. A Lean implementation requires a team to spend significantly more time in planning than they typically would on a traditionally managed project. For that reason, new teams often assign relative value to project elements, and give certain areas less attention than others. Team building and onboarding regularly receive the least attention in the planning phases and, therefore, tend to fall short in execution.

ONBOARDING

Imagine this. You are a site superintendent for masonry on a $60 million project that is employing the Last Planner System®. You are a Last Planner! You are replacing the original site superintendent that needed to leave the company due to health issues. You know that the project leadership team hosted a multi-day kick-off session, including training, team building, and an initial Pull Plan. You are excited about the challenges and opportunities of this project. When you arrive on your first day, you are given a spreadsheet and told to submit your Weekly Work Plan on Wednesday and attend the Thursday afternoon WWP Meeting. What?!?

New people will join the team throughout the project life cycle. If we want these new people to quickly become high performing members of the team, it is imperative that we take the time to plan for onboarding.

In the initial kick-off session, the project team should begin planning for initiating new team members in the future, and even consider creating a Work Cluster for onboarding. Things to consider and plan for include:

- **Full Training Session Schedule** – Depending on your project schedule, it probably makes sense to hold multiple facilitated training sessions at intervals that will capture large groups of new team members, such as transition from design to construction and start of finish activities.

- **Onboarding Agenda** – When new members join the team, time should be taken to not only introduce them to the site and explain their responsibilities within the team, but also convey the overall Lean

project approach and why things are being done the way they are. To ensure this occurs with each new addition to the team, there should be a specific agenda covering all necessary topics.

- **Facilitation Responsibility** – One strategy to achieving mastery of a new topic or process is to teach it to someone else. For this reason, several, if not all, team members should be prepared to present the initial onboarding agenda to new members, and the responsibility should rotate, instead of being assigned to one individual.

- **Ongoing Support** – Onboarding is an ongoing effort that needs to continue through project completion, in the form of coaching, upper-level support, continuing education, and team building. Providing reliable, continuous support for team members requires deliberate planning and execution.

TEAM BUILDING

Events and activities that provide team members opportunities to get to know each other outside of their roles at work, grow together as a cohesive unit, and learn from and with each other are key to creating a Lean culture, as well as creating and maintaining a high performing team. Like onboarding, planning for continuous team building should begin during the project kick-off. Further, a regular schedule of team building activities should be established early, so they don't get forgotten amid the day-to-day project activities.

Team building can, and should, take multiple forms. Here are some examples:

- **Group Outings** – Getting the team together after hours to participate in non-work-related activities is a great way to develop interpersonal relationships and break down traditional barriers between the different organizations involved in delivering the project. Perhaps consider making a different team member responsible for choosing the outing each month. Let them use their imaginations to come up with something fun, such as:

  » Sporting events

  » Happy hours

» Family activities.

» Competitive activities (ex. Bowling, skeet shooting, etc.)

- **Learning Opportunities** – Learning together is an effective way to strengthen the bond between team members.

  » **Training** – Plan to provide training throughout the project to support necessary skills, as well as simply help team members grow as professionals.

  » **Facilitated Book Studies** – Having small groups read a relevant book and meet weekly to discuss what they're learning helps to cultivate a team with common understandings and vocabularies.

- **Team Retreats** – Essentially, the project kick-off is the first team retreat. The idea is to get the team away from the day-to-day work of the project and spend time focusing on multiple topics and activities, possibly including project planning, continuous improvement, and general team building. Each retreat will take a different form, depending on the purpose and desired outcome, so put effort into developing an effective agenda to achieve the goals.

Remember that your team members are human beings that thrive on relationships with one another. When focus is put on nurturing those relationships, the project wins!

APPENDIX A

SAMPLE MEETINGS

In Chapter 4, we talked about the meetings that are a part of the Last Planner System® (Weekly Work Plan Meeting and Daily Huddle/Stand-up Meeting). Here, you'll find sample meeting scripts to help get the most out of those meetings.

This section contains sample meeting scripts for:

- WEEKLY WORK PLAN MEETING (CONSTRUCTION)
- WEEKLY WORK PLAN MEETING (DESIGN)
- DAILY HUDDLE/STAND-UP MEETING (CONSTRUCTION)
- HUDDLE/STAND-UP MEETING (DESIGN)

WEEKLY WORK PLAN MEETING (CONSTRUCTION) — 60 MINUTES

Facilitator (Jeff): Good morning, everyone! Thank you for arriving on time. Let's get started, so we can get everyone back on the job.

First on the agenda is a quick report on safety and other general information about the project. Cindy?

Sr. Superintendent (Cindy): Thanks, Jeff. It's going to be 90 degrees most of the week. Please make sure you have enough water and ice for your crews, and that you give everyone ample breaks to avoid heat exhaustion. If you need ice or water, see Jeff in the trailer. He'll get you covered.

We're doing a jobsite cleanup on Friday. Please have one person for every ten you have on site meet up at the trailer at 8:30AM to get started.

Parking will move from the West Lot to the East Lot, starting next Tuesday. We will have signage and laborers directing traffic that morning.

Any questions?

If questions can be answered easily, answer them immediately. If they require further discussion or will take you beyond the time limit for the agenda item, add them to the Parking Lot for follow-up.

Facilitator (Jeff): Let's move on to last week's plus/delta.

Most of the team seemed to appreciate the jobsite cookout that we hosted two weeks ago, so we are working to keep that going. We've talked to all of the trade partners, and they've each agreed to rotate hosting a site lunch once a month.

With the heat, we had a lot of requests for a covered area for breaks and lunch. We are looking into having a tent erected on the south side of the site. We'll give you an update next week.

We'll talk about new plus/deltas at the end of the meeting, but does anyone have any questions or comments about the actions we've taken on last week's plus/deltas?

If questions can be answered easily, answer them immediately. If they require further discussion or will take you beyond the time limit for the agenda item, add them to the Parking Lot for follow-up.

Facilitator (Jeff): Now onto how we did last week. Our Percent Plan Complete was 69%. The three biggest variances were:

Rain on Tuesday and Wednesday which delayed the sheathing and concrete formwork. Both crews worked over the weekend to make up the time.

The delivery of studs for the second floor showed up a couple days late, but

Dale (Drywall Superintendent) is sure they can still keep ahead of the in-wall rough-in.

We had to shut the site down Friday afternoon for an impromptu Owner walk-through, so we'll see if we need to meet to deal with that after today's Weekly Work Plan discussion.

Most of our variances, so far, appear to be a result of material and equipment delivery delays. We're meeting with all of the trade partner Project Managers tomorrow to come up with a solution going forward.

Any questions or comments?

If questions can be answered easily, answer them immediately. If they require further discussion or will take you beyond the time limit for the agenda item, add them to the Parking Lot for follow-up.

Facilitator (Jeff): Next, we're going to move onto look-ahead planning. Steve, can you give us an update of changes on the Constraint Log since our last weekly meeting?

Project Engineer (Steve): Absolutely, Jeff.

Since last week, we received a resolution on RFI #27 and were able to release the structural steel fabrication for the maintenance building. The Owner selected the flooring for the common area, and we received shop drawing approval for the interstitial space structure.

We also sent out change orders for Bulletin #4 on Wednesday. Unfortunately, we missed our "Need By" date by a week. We're going to have a targeted Pull Plan tomorrow morning to figure out a revised sequence for the penthouse space, so we can stay on schedule.

We're within a week of needing a resolution on RFI #29 and the submittal approval for the hardscape concrete mixes, and it looks like we won't have a problem with either of those. We'll let you know if anything changes.

Facilitator (Jeff): Thanks, Steve. Next, everyone received a revised 6-Week Look-Ahead schedule on Tuesday to review. Cindy, can you give us a rundown of how the project will look in 6 weeks? Everyone, please think about what activities you would need to complete to get to the point Cindy is going to describe, and what could possibly stop you from completing it.

Sr. Superintendent (Cindy): In 6 weeks, we will be finishing up the roof and the windows will be about 50% complete. Drywall will be hung, mudded, and sanded on the first floor, hung on the second floor, and about 25% hung on the third floor. All overhead rough-in will be complete, and in-wall will be complete through the fourth floor. EIFS will be complete up to the fourth floor. Landscapers will be mobilizing to start their fine grading and topsoil.

Facilitator (Jeff): OK, with all of that in mind, what constraints can we identify that could stop us from achieving that plan. Steve is going to record them on the Constraint log.

Electrical Superintendent (Sarah): If we're going to finish the rough-in on the fourth floor, we need a final decision on the office arrangement. To meet our pull plan commitment, we would need to know by next Thursday, so we can get our conduit prefabricated in time.

Drywall Superintendent (Dale): We'll need that, too. Thursday would be plenty of time for us.

Plumbing Superintendent (Dan): Same here, and Thursday is good for us, too.

Project Engineer (Steve): *Adds the constraint to the Constraint log with a Need By date of next Thursday.*

Who from the leadership team will make sure this constraint gets cleared?

Project Manager (Jill): I'll take care of it, and I'll get an answer by next Tuesday.

Project Engineer (Steve): *Adds Jill's name to the Champion column and next Tuesday's date for the Promised Date.*

This process repeats for all of the constraints offered by the Last Planners.

Facilitator (Jeff): Let's get into our Weekly Work Plan for next week. Sarah, we'll start with Electrical.

Electrical Superintendent (Sarah): I've got a crew working all next week on in-wall rough-in. We'll finish up the second floor on Tuesday, as long as the studs will be finished by Monday afternoon.

Drywall Superintendent (Dale): Yes, we'll have those studs up by end of the day on

Friday.

Electrical Superintendent (Sarah): Great. We'll start working on the third floor on Wednesday morning, and we'll be through all of the walls through column line R by the end of the week. My overhead rough-in crew will finish up on the fourth floor by the end of the week.

Sr. Superintendent (Cindy): Sarah, I don't see anything on your WWP about finishing up the lightning protection. The pull plan shows that taking place this week, and I know the roofers need it completed before they can finish everything up.

Electrical Superintendent (Sarah): Yes, we got word that the materials wouldn't be here until next Friday, so we're going to finish it up a week from Monday. I've spoken to Tim (Roofing Foreman), and he said that will work for them.

Sr. Superintendent (Cindy): OK, sounds good.

Facilitator (Jeff): Do you have any Workable Backlog?

Electrical Superintendent (Sarah): Yes, sorry I didn't put that on there. If we run out of work, we can move onto overhead rough-in on the fifth floor. Our light poles will be arriving on Wednesday, so we could also get started on some site lighting if we need something to do.

Throughout this conversation, the designated member of the leadership team records any changes to the team WWP.

This process repeats for each Last Planner.

When everyone has presented their WWP, the team agrees to the finalized Team WWP for the upcoming week.

Facilitator (Jeff): Finally, let's do a quick round robin for questions, new items, and plus/deltas. If we can answer questions quickly, we will. Otherwise, Steve will add them to the Parking Lot, and we'll follow-up after this meeting. Steve will also record things that you think are going well and areas where you think we can improve on the plus/delta board. I encourage you to offer any suggestions, so we can keep driving toward continuous improvement.

Give each meeting attendee an opportunity to speak.

Facilitator (Jeff): Thank you, everyone. Meeting adjourned.

Saying, "Meeting adjourned," signals the group that the WWP Meeting has ended. This eliminates any confusion when attendees proceed to carry on further discussion of Parking Lot items, and illustrates respect for the team members' time.

WEEKLY WORK PLAN MEETING (DESIGN) — 60 MINUTES

Facilitator (Maria): Good morning, everyone! Thank you for arriving on time. Let's get started, so we can get everyone back to work.

First on the agenda is a quick report on safety and other general information about the project. Joe?

Project Architect (Joe): Thanks, Maria. Just a reminder that, now that site work has begun on the project, you need to wear a hardhat and safety glasses when you visit the site.

I want to introduce you all to the new BIM coordinator on the project, Ramon. Ramon is coming straight from a $200 million hospital project, that was also employing Target Value Delivery, so he should be a great addition to the team. Welcome, Ramon!

BIM Coordinator (Ramon): Thanks! It's a pleasure to be here.

Project Architect (Joe): Finally, we're going to move our huddle meetings from Tuesday and Thursday at 1PM to the same days at 2PM, starting next week, to accommodate the new Big Room meeting schedule.

Any questions?

If questions can be answered easily, answer them immediately. If they require further discussion or will take you beyond the time limit for the agenda item, add them to the Parking Lot for follow-up.

Facilitator (Maria): Let's move on to last week's plus/delta.

As we've already mentioned, the Big Room meeting schedule was adjusted, based on feedback we received in both the Design & Construction WWP Meetings.

Besides the meeting schedule, we have received overwhelmingly positive responses about the general usage of Big Room and Work Clusters. Specifically, the team seems to appreciate the distribution of work that the Work Clusters create.

We'll talk about new plus/deltas at the end of the meeting, but does anyone have any questions or comments about the actions we've taken on last week's plus/deltas?

If questions can be answered easily, answer them immediately. If they require further discussion or will take you beyond the time limit for the agenda item, add them to the Parking Lot for follow-up.

Facilitator (Maria): Now onto how we did last week. Our Percent Plan Complete was 72%. The three biggest variances were:

Failure to lock the floor plan on Tuesday, due to not receiving a final decision on the number of offices required in the Administrative area.

Rework required on the mechanical room design, based on unexpected feedback from the Building Department. We have a call scheduled with them to figure out how quickly we can resolve the issue.

The site work contractor found an unexpected underground tank on the south side of the site. Christine (Civil Engineer) is bringing in a couple of other people to focus on this challenge, so we don't lose any time in other areas.

Most of our variances, so far, appear to be a result of not receiving answers and selections from the Owner in a timely manner. After speaking with them, we think we can come up with a more reliable process. We'll be having a team meeting with the Owner on Monday to devise a plan for going forward.

Any questions or comments?

If questions can be answered easily, answer them immediately. If they require further discussion or will take you beyond the time limit for the agenda item, add them to the Parking Lot for follow-up.

Facilitator (Maria): Next, we're going to move onto look-ahead planning. Kate, can you give us an update of changes on the Constraint Log since our last weekly meeting?

Design PM (Kate): Absolutely, Maria.

Since last week, the Core Team was able to make a final selection for the mechanical system, and we received approval from the Health Department on the kitchen design.

We finally came up with a solution for the ceiling space issues we were having. Unfortunately, we missed our "Need By" date, which is going to put our construction documents behind by about a week. We're going to have a targeted Pull Plan, next Monday, with the Construction Team to work out a plan, so we can stay on schedule.

We're within a week of needing approval of the overhead coordination for the third floor, but it looks like we won't have a problem meeting the date.

Facilitator (Maria): Thanks, Kate. Next, everyone received a revised 6-Week Look-Ahead schedule on Tuesday to review. Joe, can you give us a vision of where design will be in 6 weeks? Everyone, please think about what activities you would need to complete to get to the point Joe is going to describe, and what could possibly stop you from completing it.

Project Architect (Joe): In 6 weeks, we will have received our permit for the structural steel and building envelope. Overhead coordination will be complete through the fourth floor. Narratives and line drawings for MEPS throughout the entire building will have been sent to the trade partners for review. Initial finish sets will have been developed and in review by the Finishes Work Cluster.

Facilitator (Maria): OK, with all of that in mind, what constraints can we identify that could stop us from achieving that plan. Kate is going to record them on the Constraint Log.

Structural Engineer (Ben): We need to get a decision on the envelope by two weeks from Thursday, so we can complete our calculations and design in time to get it submitted for permit in three weeks

Project Architect (Joe): Actually, make the Need By date next Wednesday, so we can get our envelope documents done, as well. We'll make sure that decision is on the agenda for next week's Big Room sessions.

Design PM (Kate): *Adds the constraint to the Constraint Log with a Need By date of next Wednesday.

Who from the leadership team will make sure this constraint gets cleared?

Project Architect (Joe): I'll take care of it, and I'll get an answer by next Wednesday.

Design PM (Kate): *Adds Joe's name to the Champion column and next Wednesday's date for the Promised Date.

This process repeats for all of the constraints offered by the Last Planners.

Facilitator (Maria): Let's get into our Weekly Work Plan for next week. Jill, we'll start with the Plumbing Work Cluster

Plumbing Engineer (Jill): Within our Work Cluster, we'll be getting plumbing fixture recommendations ready for Core Team selection next Wednesday. We're also working on narratives and line drawings for the third floor. Those should be complete by next Friday. As far as Workable Backlog goes, if we finish up early, we'll move onto the next floors.

Design PM (Kate): Can you also get us your space needs for the main shaft by Thursday, so we can work on the final common area layout?

Plumbing Engineer (Jill): Yes, we'll get it to you on Wednesday.

Design PM (Kate): Great! Thanks.

Throughout this conversation, the designated member of the leadership team records any changes to the submitted WWP.

This process repeats for each Last Planner.

When everyone has presented their WWP, the team agrees to the finalized Team WWP for the upcoming week.

Facilitator (Maria): Finally, let's do a quick round robin for questions, new items, and plus/deltas. If we can answer questions quickly, we will. Otherwise, Kate will add them to the Parking Lot, and we'll follow-up after this meeting. Kate will also record things that you think are going well and areas where you think we can improve on the plus/delta board. I encourage you to offer any suggestions, so we can keep driving toward continuous improvement.

**Give each meeting attendee an opportunity to speak.*

Facilitator (Maria): Thank you, everyone. Meeting adjourned.

**Saying, "Meeting adjourned," signals the group that the WWP Meeting has ended. This eliminates any confusion when attendees proceed to carry on further discussion of Parking Lot items, and illustrates respect for team members' time.*

DAILY HUDDLE/STAND-UP MEETING (CONSTRUCTION) — 15-20 MINUTES

Sr. Superintendent (Cindy): Thanks for being on time, everyone. Let's get started.

Since yesterday, we received an answer for RFI #29. You should have all received it, and we sent it to your PM's as well.

Anyone have any constraints that we should add to the log.

Plumbing Superintendent (Dan): We'll need the rebar moved by next Friday, so we can finish up the site plumbing the following week.

Concrete Foreman (Bryan): We can do that.

Sr. Superintendent (Cindy): OK, I'll add it to the Constraint Log.

Anything else?……. Alright, is everyone on track to finish today's plan on the WWP, and move onto tomorrow's work as planned? Does anyone need anything from anyone?

Concrete Foreman (Bryan): We're good to have the elevated slab prepped by the end of the day, but we've moved tomorrow's pour back a day, because it looks like rain all day tomorrow. We're going to spend the day prepping the housekeeping pads in the mechanical space.

Electrical Superintendent (Sarah): Hey, Bryan. Can you start on the east side of the space and move west? We've got some overhead work that spans a couple of those housekeeping pads. We should be finished by lunch, though.

Concrete Foreman (Bryan): No problem.

This continues with brief updates from each of the Last Planners.

Cindy makes note of variances and their reasons on the Team WWP for input to the Percent Plan Complete and Variance Log.

Sr. Superintendent (Cindy): Thanks, everyone. Meeting adjourned.

Saying, "Meeting adjourned," signals the group that the meeting has ended. This eliminates any confusion when attendees proceed to carry on further discussion, and illustrates respect for team members' time.

HUDDLE/STAND-UP MEETING (DESIGN) — 15-20 MINUTES
Design Huddles often occur two or three times a week, rather than daily like Construction Huddles.

Facilitator (Maria): Thanks for being on time, everyone. Let's get started.

We haven't had any constraints cleared since our last meeting, but we're expecting to have several to update in a couple of days. Nothing is expected to be late at the moment.

Anyone have any constraints that we should add to the log.

Project Architect (Joe): Yes, can you add that we need a Core Team selection on whether or not there will be a suspended running track in the fitness center in the next 5 weeks, so we can finish the design in time to submit for permit?

Facilitator (Maria): Sure. Are you going to champion that item, Joe?

Project Architect (Joe): Yes, I'll make sure we get a decision by the 27th.

Facilitator (Maria): Great!

Anything else?....... Alright, is everyone on track to meet this week's deadlines on the WWP? Does anyone need anything from anyone?

Plumbing Engineer (Jill): We're supposed to have the plumbing fixture recommendations to the Core Team by Tuesday for selection on Wednesday. We just got word that the Facilities Management Department is making some specific requests that we hadn't taken into consideration. We're meeting with them on Tuesday, so the earliest we can get a revised analysis and recommendation ready is Wednesday. Can we move it to later in the Big Room agenda? If it looks like we won't make it, we'll let you know Tuesday afternoon.

Project Architect (Joe): OK. Keep us informed, and let us know if you need any assistance or if we need to meet to adjust the Pull Plan.

This continues with brief updates from each of the Last Planners.

Maria makes note of variances and their reasons on the Team WWP for input to the Percent Plan Complete and Variance Log.

Facilitator (Maria): Thanks, everyone. Meeting adjourned.

Saying, "Meeting adjourned," signals the group that the meeting has ended. This eliminates any confusion when attendees proceed to carry on further discussion, and illustrates respect for team members' time.

"I ALREADY DO THIS. WHY DO I NEED LEAN?" AND OTHER FREQUENTLY ASKED QUESTIONS

I've been employing, facilitating, teaching, and championing Lean Design & Construction philosophies and Last Planner System® tools for the past twenty years. From the moment that I began leading my first Lean project, I knew that this was the best way to manage work. Our simple use of Last Planner tools, such as Pull Planning, Weekly Work Plans, and Look-Ahead Plans, created an atmosphere of teamwork and planning that I had never experienced on a project before. Since that first project, I've learned more about Lean, brought its concepts to multiple organizations, and, ultimately, made supporting organizations and project teams along their Lean journeys the focal point of my career.

Over the last two decades, I've been asked a lot of questions about Lean — some out of genuine curiosity and a desire to learn — others in an attempt to undermine a Lean effort. Both are fine with me. I appreciate the discussion, and it's kind of fun when a nay-sayer suddenly sees the light!

Of course, I've had years of practice considering these queries and honing my responses. I know there are Lean champions out there working to build support for a Lean transformation within their companies that haven't yet formulated their answers to all the questions and challenges that they are sure to face. So, I thought I would share some Frequently Asked Questions about Lean Design & Construction, along with my answers. I hope they help.

ARE LEAN DESIGN & CONSTRUCTION AND THE LAST PLANNER SYSTEM THE SAME THING?

No. Lean Design & Construction is a management philosophy in which we strive to create a collaborative environment, aimed at reducing all forms of waste and nurturing a culture of Continuous Improvement. The Last Planner System, although an integral part of a Lean project, is a group of management and planning tools (Pull Plan, Weekly Work Plan, Look-Ahead Plan, Percent Plan Complete, Daily Huddle, etc.) that we implement on our projects to foster Lean thinking and support effective project planning and communication. These tools harness the collective experience, knowledge, and authority of the design leads, project foremen, and superintendents, or "Last Planners," to create and execute a reliable and flexible project plan.

WHY DO I NEED LEAN, IF I'M ALREADY COLLABORATIVE AND PLANNING AHEAD? (VARIATION — "I ALREADY DO THIS.")

During my many years in commercial construction, I have met several truly exceptional construction managers that seek out the input of their trade partners for project planning, as well as problem identification and solution. I have never, however, been on a project populated only by such individuals. A natural tendency toward teambuilding and working together is just that, natural. Without any effort to develop this ability, it's either a personality trait that you have, or one that you don't. Implementing Lean, and specifically the Last Planner System tools, provides a step-by-step guide to participating in a collaborative environment, even if it's not in your nature. Remember, practice makes perfect, and when value can be seen from new activities, it's habit-forming. Suddenly that forward-thinking, team-focused leader is surrounded, and supported, by a project team that are all working toward the same goals.

WHERE DO THE SAVINGS COME FROM IN LEAN?

Cost savings (and financial gains) from implementing Lean on your projects comes in many forms. Some you'll realize right away. Others come a little later.

Short-Term Savings: Your very first Lean project will experience the cost benefits that come from improved planning and communication, as well as your team's efforts to develop and maintain a reliable flow of work. When the different team members feel like they're "in this together," they work together to solve problems before they ever become an issue for the project. That alone results in fewer change order claims and schedule delays. Your schedule, built on reliable commitments from the Last Planners, is more manageable, and more adaptable to unforeseen issues and changes. That translates into shorter overall durations (lower general conditions and overhead costs), less overtime, and fewer last-minute expenditures that may be needed to "make the date."

Long-Term Savings (and Gains): As time passes, and designers and trade partners realize that they are saving money when they work in the Lean environment that you've created, that savings will be reflected in the proposals that they submit to your organization. In turn, you will be able to propose lower prices than your non-Lean counterparts, resulting in more project awards. Further, the innovations that are borne of your culture of Continuous Improvement will garner savings through better efficiency, higher productivity, further shortened project durations, and constantly improving teamwork on your projects.

HOW CAN WE EXPLAIN THE BENEFITS OF PARTICIPATING IN A LEAN PROGRAM TO OUR TRADE PARTNERS AND DESIGN CONSULTANTS?

Often, the initial Lean and Last Planner System implementation is met with claims that the processes and planning are too much work and a drain on resources. After all, you're asking your trade and design partners to send their Last Planners to boot camps and Pull Plan sessions, reliably plan their work for each week down to the day, spend time communicating with the other project team members, attend daily huddles, and, in general, focus on the good of the project, instead of their own bottom line. Why would they ever want to do that?

The fact is, Lean wouldn't work, and it certainly wouldn't have lasted as long as it has, if it wasn't good for everyone involved. Achieving a reliable flow of work throughout the duration of a project increases productivity and worker morale, reduces stress and animosity, and shortens the overall duration. Shorter durations result in lower general conditions and overhead costs, less overtime and last-minute scrambling, and a happier client that is inclined to favor this project team the next time they plan to build or renovate.

Additionally, since Lean is here to stay, any organization that can show experience and expertise in its processes and implementation can differentiate themselves in future project proposals. General Contractors and Construction Managers spend valuable resources to learn about Lean, train their employees, and coach pilot project teams to ensure a successful roll-out within their organizations. Project partners get many of the same benefits simply by being part of the project team and participating fully.

IF WE ONLY DO PULL PLANNING, ARE WE STILL RUNNING A LEAN PROJECT?

Pull Planning is a great first step in any organization's Lean journey. The simple act of bringing all of your Last Planners together to collaboratively plan a project puts your team ahead of many on traditionally managed projects. However, simply building a Pull Plan before each phase of work does not mean you're running a Lean project. Remember that Lean seeks to eliminate waste and continuously improve the overall process. The Pull Plan alone cannot do this. To get all of the benefits of Lean, you need to get into a cycle of PLAN-DO-CHECK-ADJUST. In other words, set out your plan (i.e. Your Pull Plan, a new on-site process, a new management innovation, etc.), implement your plan, look back to judge the effectiveness of what you're doing, and make necessary adjustments. When you only do the Pull Plan, and manage traditionally from there, you're missing several of the steps necessary to be truly Lean.

Additionally, most companies begin their Lean journey by implementing the Last Planner System. This is perfect, but you shouldn't stop there. As your teams grow in their ability to seek out new challenges and work together to solve them, it's time to stretch into more advanced Lean thinking – Enhanced Visualization, Target Value Delivery, Big Room Methodology, Integrated Project Delivery (IPD), and more.

GLOSSARY

Courtesy of the Lean Construction Institute

5S - A disciplined approach to maintaining order in the workplace, u s i n g visual controls, to eliminate waste. The 5S words are Sort, Set in Order, Shine/Sweep, Standardize, and Self- Discipline/Sustain.

5-Why Analysis - The problem-solving technique used to dig for the root cause of a condition by asking why successively (at least five times) whenever a problem exists, in order to get beyond the apparent symptoms. As each answer to the why question is documented, an additional inquiry is made concerning that response.

A3 - A one-page report prepared on a single 11 x 17 sheet of paper that adheres to the discipline of PDCA thinking as applied to collaborative problem solving, strategy development, or reporting. The A3 includes the background, problem statement, analysis, proposed actions, and the expected results.

Activity - An identifiable chunk of work with recognized prerequisite requirements to begin and a recognized state of completion - or conditions of satisfaction. Another way to look at an activity - establish the hand-offs for each chunk of work thus defining the activity. (see also "task")

Actual Cost - The sum of the total Cost of the Work actually incurred by Architect and CM/GC in connection with the performance of all Phases of the Project, plus CM/GC's Fee. (IFOA - Integrated Form of Agreement Definition)

Allowable Cost - The absolute maximum Project Cost, based on the Project Business Case as outlined in Exhibit 2, which will be the subject of the Validation Study. (IFOA - Integrated Form of Agreement Definition)

Assignment - A request or offer that has resulted in a Reliable Promise and is ready to be placed on the Weekly Work Plan for performance. An assignment must meet the characteristics for a Quality Assignment prior to inclusion on the WWP.

Buffer - As a verb: "to isolate one activity from the next." A mechanism for deadening the force of reality unfolding in a manner that is contrary to what was anticipated in the plan. For example, a capacity buffer is created by committing to complete less work than what would be achieved according to the planned capacity of the resource. If production falls behind schedule, there is capacity available for catching up. (Lean production/construction generally prefers capacity buffers to inventory buffers.)

Building Information Model(-ing) (BIM) - The process of generating and managing building data during the life cycle of a building. BIM uses three-dimensional (3D), real-time, dynamic building modeling software. BIM includes building geometry, spatial relationships, geographic information and quantities, and properties of building components. BIM can include four- dimensional (4D) simulations to see how part, or all, of the facility is intended to be built and 5D capability for model-based estimating. BIM provides the platform for simultaneous conversations related to the design of the "product" and its delivery process.

Capacity - The amount of work that can be produced by an individual specialist or work group in a given period of time.

Choosing by Advantages (CBA) - CBA is a tested and effective sound decision-making system developed by Jim Suhr (1999) for determining the best decision by looking at the advantages of each option. CBA's five phases of decision-making:

1. **Stage-setting**: *establish the purpose and context for the decision;*

2. **Innovation**: *formulate an adequate set of alternatives;*

3. **Decision-making**: *choose the alternative with the greatest total importance of advantages;*

4. **Reconsideration**: *change the decision if it should be changed or improved on;*

5. **Implementation**: *make the decision happen, adjust as needed, and evaluate the process and results.*

Commitment Based Planning - A planning system that is based on making and securing reliable promises in a team setting.

Conditions of Satisfaction - An explicit description by a Customer of all the actual requirements that must be satisfied by the Performer in order for the Customer to feel that he or she received exactly what was wanted.

ConsensusDocs - A library of more than 100 standard contract documents written and endorsed by a coalition of more than 40 leading design and construction industry organizations, including LCI. The ConsensusDocs 300 Agreement is a standard IFOA that incorporates Lean practices in the agreement. (see also "IFOA")

Constraint - An item or requirement that will prevent an activity from starting, advancing, or completing as planned. Typical constraints on design tasks are inputs from others, clarity of requirements criteria for what is to be produced or provided, approvals or releases, and labor or equipment resources. Typical constraints on construction tasks are the completion of design or prerequisite work, availability of materials, information, and directives. Screening tasks for readiness is assessing the status of their constraints. Removing constraints is making a task ready to be assigned.

Constraint Log - A list of Constraints with identification of an individual promising to resolve the item by an agreed date. Typically developed during a review of the Six Week Look-Ahead Plan when it is discovered that activities are not constraint free.

Cost Modeling - Developing a model of the cost components and systems specific to a project and structuring it in a manner that the components and system costs can be continually updated either via benchmarks, metrics, or detailed estimates to provide the team with a constantly up-to-date cost model for the project. In the TVD environment, the cost model should allow for projecting 'what-if' scenarios based on value decisions that have yet to be made.

Customer - The individual engaged in a conversation for action who will receive the results of performance either requested from, or offered by, the performer. i.e. – the person receiving goods/information from a performer. Customers can be internal (a foreman receiving answer to an RFI, Architect receiving mechanical loads from engineer), and external (end users, client organizations, etc.).

Cycle Time - The time it takes a product or unit of work (e.g. a room, building, quadrant) to go from beginning to completion of a production process; i.e., the time it is work-in-process.

Defined Task - A Quality Task must be "Defined." It must have a beginning and end. It should be clear to all when it has been completed.

Dependence - Where two or more tasks are sufficiently related that one cannot be started (or finished) without a certain measure of progress or completion having been achieved by the other. Waiting on release of work.

Expected Cost - An expression of the team's best estimate at the conclusion of the Validation Phase of what current best practice would produce as a price for the facility reflected in the accompanying basis of design documents. Typically, the Expected Cost will also be supported by benchmarking or other market data to calibrate the Expected Cost in light of the market context.

First-run Study - Trial execution of a process in order to determine the best means, methods, sequencing, etc. to perform it. A FRS follows the Plan-Do-Check-Adjust cycle. First-run studies are done at least a few weeks ahead of the scheduled execution of the process, while there is time to acquire different or additional prerequisites and resources. They may also be performed during design as a basis for evaluating options or designing the portion of the work.

Five Big Ideas - A set of organizing concepts that support Lean Project Delivery. They were developed to explain and organize the Sutter Health Lean Construction Initiative: Optimize the project not the piece, Collaborate, Really Collaborate (originally implied "specialty contractors involved at schematic design"), Projects as Networks of Commitment, Increase Relatedness, and Tightly Couple Action and Learning.

Flow - Movement that is smooth and uninterrupted, as in the "flow of work from one crew to the next" or the flow of value at the Pull of the customer.

Gemba - The Japanese term for where value is added or where the work takes place. Lean experts encourage "going to the gemba" to see how things are really done and where there is opportunity to eliminate or reduce waste.

Hand-off - The act of releasing an item or activity to the person or group performing the next step or operation on that item or activity. Example: a structural steel design is "handed off" to the steel detailer to complete shop drawings; a room (or portion) that has been framed is "handed off" to the drywall installer; or all construction on a floor of a hospital is completed and it is "handed off" to the hospital personnel to begin staff-and-stock activities.

Hand-off Criteria - The Conditions of Satisfaction discussed and explicitly agreed upon between the parties to a hand-off.

Integrated Form of Agreement (IFoA) - A multi-party agreement that includes the owner, design professional, and constructor as signatories to the same construction contract.

Integrated Project Delivery (IPD) - A project delivery approach that integrates people, systems, business structures and practices into a process that collaboratively harnesses the talents and insights of all participants to reduce waste and optimize efficiency through all phases of the project, from early design through project handover. The three contractual components of IPD include Organization structure, Lean Operating Systems and Commercial Terms.

Just-in-Time - A system for producing or delivering the right amount of parts or product at the time it is needed for production. ("JIT")

Kaizen - The Japanese word for continuous improvement. Kaizen has come to mean the philosophy of continuous improvement.

Kanban - Japanese term meaning "a signboard." A communication tool used in JIT production systems. The signal tells workers to pull parts or refill material to a certain quantity used in production.

Last Planner® - The person or group that makes assignments to direct workers. Project Architect and 'discipline lead' are common names for last planners in design processes. 'Superintendent' or 'foremen' are common names for last planners in construction processes.

Last Planner System® (LPS) - System for project production planning and control, aimed at creating a workflow that achieves reliable execution, developed by Glenn Ballard and Greg Howell, with documentation by Ballard in 2000. LPS is the collaborative, commitment-based planning system that integrates should-can-will-did planning: pull planning, make-ready look-ahead planning with constraint analysis, weekly work planning based upon reliable promises, and learning based upon analysis of PPC and Reasons for Variance.

Last Responsible Moment (LRM) - The instant in which the cost of the delay of a decision surpasses the benefit of delay; or the moment when failing to make a decision eliminates an important alternative.

Lean - Culture of respect and continuous improvement aimed at creating more value for the customer while identifying and eliminating waste.

Lean Project Delivery System - An organized implementation of Lean Principles and Tools combined to allow a team to operate in unison to create flow.

Load - The amount of output expected from a production unit or individual worker within a given time.

Look-Ahead Planning - The portion of the Last Planner System® that focuses on making work ready - assuring that work that should be done, can be done, by identifying and removing constraints in advance of need.

Look-Ahead Plan - A short interval plan, based on the pull/phase plan, that identifies all the activities to be performed in the next 6 (or other) weeks. The 6W Look-ahead Schedule (LAS) is updated each week - always identifying new activities coming 6 weeks out so that the project management team can make appropriate arrangements to assure that the work will be ready to be performed in the week indicated.

Look-Ahead Window - The duration associated with Look-Ahead Planning. Typically look-ahead windows extend from 3 to 12 weeks into the future, with six weeks preferred on most projects.

Make Ready Process - To "make ready" is to take actions needed to remove constraints from assignments to ensure the work can be done as planned.

Master Schedule - A schedule that identifies major events or milestones in a project (start- up, turn-over to client, order long delivery components, mobilize in field, complete design, government reviews, etc.) and their timing. It is often the basis for contractual agreements between the owner and other team members. It is seen as a way to identify long lead items, the feasibility of completing the project as currently required, the basis for defining milestones and phases – but not always as a way to "control" the project.

Milestone - An item on the Master Schedule that defines the end or beginning of a phase or a contractually required event.

Muda - Japanese word for "Non-value-added" or Ohno's 7 Wastes.

Mura - Japanese word for "Unevenness" - fluctuation in demand that causes the workflow to be uneven.

Muri - Japanese word for "Overburdening" - excessive demand on a system that causes the system to produce beyond its reasonable capacity. Pushing a machine or person beyond natural limits. Overburdening people results in safety and quality problems. Overburdening equipment causes breakdowns and defects.

Network of Commitments - The web of promises necessary to deliver any project. The role of management is to articulate and activate the unique network of commitments required to deliver each project.

PDCA - Stands for Plan - Do - Check - Adjust. The cycle introduced by Walter A. Shewhart and popularized by Dr. W. E. Deming as a method of continuous improvement.

Percent Plan Complete (PPC) - A basic measure of how well the planning system is working - calculated as the "number of promises/activities completed on the day stated" divided by the "total number of promises/activities made/planned for the week." It measures the percentage of assignments that are 100% complete as planned.

Performer - The individual engaged in a conversation for action who agrees to undertake performance either requested from or offered to a Customer.

Phase - A period of the project where a specific group of activities is scheduled to be accomplished such as building design, completion of foundations, erection of exterior walls, building dry-in, etc. A phase can be either a time period or a group of activities leading to the accomplishment of a defined goal/milestone.

Phase Plan or Pull Plan - A plan for executing a specific phase of a project using a pull technique to determine hand-offs. It is prepared by the team actually responsible for doing the work through conversation. Work is planned at the "request" of a downstream "customer."

Plan Reliability - The extent to which a plan is an accurate forecast of future events, measured by Percent Plan Complete (PPC).

Planning - The act of conversation that leads to well-coordinated action.

Plus/Delta Review - A continuous improvement discussion preformed at the end of a meeting, project, or event and used to evaluate the session or activity. Two questions are asked and discussed. Plus: What produced value during the session? Delta: What could we change to improve the process or outcome?

Poke yoke - A Japanese term for a mistake-proofing method or device developed by Shigeo Shingo that is used to prevent an error or defect from happening or being passed on to the next operation.

PPE - Personal Protective Equipment, commonly referred to as "PPE", is equipment worn to minimize exposure to serious workplace injuries and illnesses.

Prerequisite work - Work that must be performed by others in order for you to perform your work.

Process Map - A flowchart identifying all the activities, operations, steps, and work times for a process.

Promise - The action taken by "Performer" to commit to a "Customer" to take some action to produce a mutually understood result ("Conditions of Satisfaction") by a definite time in the future. (see also "Reliable Promise")

"Pull" - A method of advancing work when the next in line customer is ready to use it. A "Request" from the customer signals that the work is needed and is "pulled" from the performer. Pull releases work when the system is ready to use it.

"Push" - "Push" - an "Order" from a central authority based on a schedule; advancing work based on central schedule. Releasing materials, information, or directives possibly according to a plan but irrespective of whether or not the downstream process is ready to process them.

Quality - Conformance to a Customer's valid and agreed upon Conditions of Satisfaction.

Quality Assignment – An assignment that meets quality criteria for release to the customer process. The quality criteria are: (1) definition, (2) soundness, (3) sequence, (4) size, and (5) learning.

Reason for Variance - Factors that prevented an assignment from being completed as promised, used by the team to promote learning concerning the failure of the planning system to produce predictable workflow. By assigning a category of variance to each uncompleted task, a team is able to identify those areas of recurring failure that require additional reflection and analysis.

Reliable Promise - A promise made by a performer only after self-assuring that the promisor (1) is competent or has access to the competence (both skill and wherewithal), (2) has estimated the amount of time the task will take, (3) has blocked all time needed to perform, (4) is freely committing and is not privately doubting ability to achieve the outcome, and (5) is prepared to accept any upset that may result from failure to deliver as promised.

Request - The action taken by a Customer" to ask a "Performer" to take some action to produce a mutually understood result ("Conditions of Satisfaction") by a definite time in the future.

Root Cause Analysis - A systematic method of analyzing possible causes to determine the root cause of a problem. (see also "5 Why Analysis")

Screening - Determining the status of tasks in the look-ahead window relative to their constraints and choosing to advance or retard tasks based on their constraint status and the probability of removing constraints.

Set-Based Design (SBD) - A design method whereby sets of alternative solutions to parts of the problem are kept open until their Last Responsible Moment(s), in order to find by means of set intersection the best combination that solves the problem as a whole.

Sequenced - A "sequenced" assignment should release work to another Performer and in no case should it hinder another assignment or cause other crews to do additional work. Quality criterion for selecting assignments among those that are sound in priority order and in constructability order.

Shielding - Preventing the release work to production units because it does not meet quality criteria; the work is not a quality assignment. It is akin to "stopping the assembly line," rather than advancing a defective product. The purpose of shielding is to reduce uncertainty and variation, thereby providing production units with greater opportunity to be reliable.

Should-Can-Will-Did - To be effective, production management systems must tell us what we should do and what we can do, so that we can decide what we will do, then compare with what we did to improve our planning.

Sized - Quality criterion for assignments whereby the amount of work included in an assignment is made to match the capacity of the production unit that will do the work. The Performer should have a very reasonable expectation that the assignment can be completed by the number of people available to do the job.

Sound - Quality criterion for assignments that tests whether or not assignments have had all constraints removed. The Performer of an assignment should know that the materials, tools, staff, and information to complete an assignment are available before accepting it.

Target Cost - The cost goal established by the delivery team as the "target" for its design and delivery efforts. The Target Cost should be set lower than best-in-class past performance. The goal is to create a sense of necessity to drive innovation and waste reduction into the design and construction process.

Target Value Delivery - A disciplined management practice to be used throughout the project to assure that the facility meets the operational needs and values of the users, is delivered within the allowable budget, and promotes innovation throughout the process to increase value and eliminate waste (time, money, human effort).

Target Value Design - Encompasses the Target Value Delivery approaches implemented during the design delivery phases of the project.

Target Value Production - Encompasses the Target Value Delivery approaches implemented during the construction delivery phases of the project.

Task - An identifiable chunk of work. (see also "Activity")

Throughput - The output rate of a production process.

Under-loading - Making assignments to a production unit, or a resource within a production unit, that absorbs less than 100% of its capacity. Under-loading is necessary to accommodate variation in processing time or production rate, in order to assure plan reliability. Under-loading is also done to release time for workers to take part in training or learning, conducting first-run studies, implementing process improvements, or for equipment to be maintained.

Utilization - The percentage of a resource's capacity that is used in actual production.

Value - What the Customer wants from the process. The customer defines value.

Value Stream - The sequence of activities required to design, produce, and deliver a good or service to a customer, and it includes the dual flows of information and material.

Value Stream Mapping - A team-based methodology for analyzing the current state and designing a future state for a series of events that take a product or service from its beginning through to the customer.

Variance - When an assignment is not completed as stated, it is considered a variance from the weekly work plan.

Visual Management - Placing tools, parts, production activities, plans, schedules, measures, and performance indicators in plain view, This assures that the status of the system can be understood at a glance by everyone involved and actions taken locally in support of system objectives.

Waste - The opposite of value. There are seven basic types of waste including: defects, waiting, transportation of goods, motion, inventory, overproduction, and unnecessary process steps.

Weekly Work Plan - The commitment-level ("will") planning step of LPS® identifying the promised task completions agreed upon by the Performers. The WWP is used to determine the success of the planning effort and to determine what factors limit performance. It is a more detailed level than the look-ahead plan and is the basis of measuring PPC (Percent Plan

Complete).

Weekly Work Planning - The process by which the Last Planners establish the plan for the coming period.

Work Flow - The movement of information and materials through networks of interdependent specialists.

Work Structuring - Designing the production system to determine who does what, when, where, and how, usually by breaking work into pieces, where pieces will likely be different from one production unit to the next. The purpose of work structuring is to promote flow and optimize system throughput by focusing on handoffs and opportunities for moving smaller batches of work though the production system.

Workable Backlog - An activity or assignment that is ready to be performed, but is not assigned to be performed during the active week in the WWP. If the team agrees that performance of this activity will not hinder other work then it can be placed on the list of Workable Backlog as part of the WWP. Completion or non-completion of these activities are not recorded or counted in calculation of PPC.

Work In Process (WIP) - The inventory between the start and end points of a production process.

NOTES